D0327111

Steel in the United States: Restructuring to Compete

Steel in the United States: Restructuring to Compete

William T. Hogan, S.J.
Fordham University

LexingtonBooks
D.C. Heath and Company
Lexington, Massachusetts
Toronto

Library of Congress Cataloging in Publication Data

Hogan, William Thomas, 1919–
Steel in the United States: Restructuring to compete

Includes index.
1. Steel industry and trade—United States. I. Title.
HD9515.H62 1984 338.4′7669142′0973 83–49519
ISBN 0–669–08234–1 (alk. paper)

Second printing, December 1984

Published simultaneously in Canada

Printed in the United States of America on acid-free paper

International Standard Book Number: 0–669–08234–1

Library of Congress Catalog Card Number: 83–49519

Contents

Acknowledgment

The author wishes to express his appreciation and acknowledge the generous cooperation, assistance, and support received from the Association of Iron and Steel Engineers. This organization, headquartered in Pittsburgh, Pennsylvania, is an international technical society, dedicated to the advancement of the technical and engineering phases of the production and processing of iron and steel.

Part I
The Integrated Steel Companies: Current Status and Future Plans

1 Introduction

The deep depression that gripped the steel industry in late 1981, and which continued into early 1984, generated a number of problems concerning the industry's future. They stemmed in great part from the large financial losses sustained by every integrated steel producer during much of this period resulting in plant closings, massive layoffs, and sharp cuts in cash flow. All of these have had a severe impact and forced virtually every major integrated company to rethink its plans for the future. In many instances, dramatic decisions have been made involving the elimination of plants, the abandonment of high-cost, obsolete facilities, a general reassessment of market approach, and diversification into nonsteel activities.

Projects that were planned in many instances have been postponed and, in a number of cases, cancelled. For example, the new integrated steel mill that United States Steel Corporation (USS) proposed to build at Conneaut, Ohio, on Lake Erie, was first indefinitely postponed and finally abandoned. This was to have been a 3-million-ton-per-year plant in its first stage and later was to be expanded considerably. The abandonment was the result of excessively high costs involved in the construction of the plant which, coupled with the relatively low prices that would be obtained for its product, would have made it a losing proposition. Another abandoned project was the integration of National Steel's Midwest plant near Chicago. The plan was to bring its finishing facilities up to full integration through the construction of coke ovens, a blast furnace, steelmaking equipment, continuous casting, and a hot-strip mill. Two years after the project was announced, a series of changes, including increased costs, a drop in cash flow, and a declining market for steel, forced a postponement and finally a cancellation. Other projects of a smaller scope (including the construction of new blast furnaces and finishing mills) have been cancelled as the investment cost increased, the market turned down, and companies reassessed the future. Currently, only those projects that are considered absolutely necessary (such as continuous-casting units) are being undertaken, and this approach has halted the progress that many companies were making toward complete modernization of their steelmaking and -finishing facilities.

As a consequence of recent actions taken by a number of integrated steel producers, the industry as of early 1985 will be considerably different from what it was in late 1981. It will be smaller in terms of raw-steel capacity, and it will be more efficient with the closure of obsolete, high-cost facilities, coupled with the installation of new equipment. In addition, decisions to engage in nonsteel activities will make the industry more diversified. As a result of the merger between Republic and J&L, there will be fewer companies.

In the light of these developments, serious questions have been raised in regard to the industry's future. These include its ability to (1) operate profitably; (2) compete on a world basis; (3) serve the needs of the U.S. market with its reduced capacity; and (4) raise sufficient capital to make further necessary facility replacements.

Generally speaking, there are two basic schools of thought on the future of the U.S. steel industry. One is basically pessimistic; the other is mildly optimistic. The first maintains that the industry (1) has failed to modernize its facilities to compete on a world basis; (2) is currently operating a considerable percentage of high-cost, obsolete equipment; and (3) requires a substantial investment in steel plants and equipment for the companies to catch up and become competitive.

In respect to the third observation, it has been maintained that the investment needs are so great that the companies will not be able to meet them completely and thus will continue to decline. This view brings into focus the problem of the renewal and modernization of the steel industry's facilities, a problem that demands careful analysis and serious attention.

The process of capital renewal has been particularly difficult for steel. Its depreciation accruals have been woefully inadequate since most of its assets carry replacement costs at least double (and often triple) their installation costs, the latter being all that depreciation provides. Not only have advances in steel technology required a replacement of assets well before the expiration of their depreciable lives, but mandated requirements for environmental control have added to their complexity and cost, as have the demands for increasing product quality imposed by the marketplace. Rarely have steel assets been replaced in kind, and virtually never was the cost covered by accumulated depreciation accruals. As a result, the steel industry has had to draw heavily on retained earnings for asset renewal. Because its earnings have been limited (and in 1982 and 1983 nonexistent) debt has been utilized to the detriment of individual company balance sheets.

The end result has been an inadequate modernization and replacement of steel facilities and a vicious circle of obsolescence that has seen steel managements forced to pick and choose between their plants in allocating limited capital, placing their available funds where the prospective returns are best, and depriving those plants that need to be modernized the most. One by one, the deprived units have been permanently shut down over the last few years, thereby shrinking the industry's productive capacity.

In considering whether or not adequate capital renewal can ever be achieved in the steel industry, it is important to recognize that steel-facility modernization and replacement no longer represent a guarantee of steel-company competitiveness. Although higher degrees of modernization are often cited to explain the competitive advantage that foreign steel producers have been able to gain in the U.S. market, a number of other, sometimes noneconomic, factors have been overriding the influence of costs and productive efficiency in determining relative steel prices and competition. In recent years, government subsidies to steel companies, trade restrictions, and below-cost pricing have been far more significant than plant and equipment modernization in establishing the U.S. steel market's competitive framework. In fact, some domestic steel companies that have been among the most progressive in adopting new technology recently have had greater financial difficulties than their less progressive counterparts, underscoring the fact that modern production facilities do not necessarily assure success in the steel business. So, a logical question for steel companies to ask is, "To what extent can modernization solve their problem?"

Positive decisions to make steel investments are made even more difficult by the extremely high cost of installing new facilities, which often are priced in the hundreds of millions of dollars. The need to borrow much of these massive funding requirements places further obstacles in the way of modernization. However modern, the new facilities will produce products that may well be sold under conditions of cutthroat competition, and possibly at a loss and most likely not at an acceptable profit. This provides a strong discouragement to both steel-company borrowers and potential lenders in the investment community. In some cases, the sheer size of the borrowing costs incurred can more than offset the potential savings in operating costs per ton that the new facilities can provide.

Another major obstacle to steel modernization derives from recent changes in the U.S. economy with respect to the rate of inflation and the level of real interest rates. Recent economic developments have reversed former trends and tend to reward savers and penalize borrowers—the former receiving more from the latter, who must now pay back the funds they borrow in real rather than sharply deflated dollars. While nominal interest rates have declined considerably over the last two years, the rate of inflation has declined even more, so that measured on a real basis, loanable funds are now more expensive to acquire, which makes debt-financed projects all the more difficult to justify. Marginal and subpar investments (which many steel projects must be considered) are less likely to be financed now that debt repayment can no longer be made in significantly cheaper dollars.

The stricter rules that must now be applied in judging the feasibility of debt financing represent just one aspect of a generally more difficult climate for steel investment. Other problems currently deterring adequate steel-facility modernization pose a serious investment dilemma, since they explain precisely why increased modernization is sorely needed. These problems include:

1. The current structure of steel prices, which is indicative of steel being marketed more and more as a commodity, with sharp discounting from list in the U.S. market and even sharper discounts prevailing in markets overseas.
2. The increasing proportion of overseas steel industries that are either government owned or subsidized, which has made many of them tools of national economic policy, divorcing their output from demand and their pricing policies from cost consideration.
3. The wide margin of labor-cost disadvantage faced by U.S. steel companies in comparison to their foreign counterparts, even with the concessions made in recent labor agreements.
4. The belief among steel managements, both in this country and abroad, that a fundamental, moderating shift in steel demand has occurred in response to the oil-price shocks of 1973–1974 and 1979–1980 and that most, if not all, steel producers in the industrialized world face a future of either minimal or no steel-market growth.
5. The new competitive threat posed by highly efficient steel producers in developing-world countries, where capacity in excess of present domestic requirements has been or is soon to be brought on stream.
6. The widening scope of competition from substitute materials such as plastics, aluminum, and cement, which have made notable inroads into steel's automotive, container, and construction markets.
7. The sheer size of the modernization requirements that steel companies face, encompassing plant and equipment throughout a wide range of their operations and involving capital costs in the hundreds of millions, if not billions, of dollars on an individual-company basis.

In combination, all of these problems have produced a pessimistic view of the steel outlook, which has become more apparent and widespread not only among the industry's critics but also among some major, integrated producers. Their independently derived views of the future, formed on the basis of their individual-company problems, have been manifested in cuts in steelmaking capacity, the permanent shutdown of major plants, the dismissal by the thousands of production workers and management personnel, the indefinite postponement and cancellation of major investment projects, and in some cases the desire to quit certain segments of the steel business. Some steel industry analysts have expressed the wiew that in the United States and some Western European countries, the industry is a dying dinosaur; they see the future mainly in the Third World, where new facilities and low labor costs can produce good, cheap steel.

There is a contrasting view that, while recognizing the problems that the industry faces, takes a much less pesimistic attitude toward its future that even borders on long-range optimism. This view takes into account the recent changes that many steel companies have made, are making, and will continue to make. It maintains that the recent steel crisis has given rise to fewer, smaller, but stronger, more efficient producing companies. This is manifested by the merger recently arranged between J&L and Republic and the attempted merger between USS and National.

The optimistic view, which seems more tenable with the steel market improving, maintains that in the future the restructured companies will be able to operate at higher rates of capacity and lower breakeven levels, so that in late 1984 and beyond, these companies should be able to show profits. Thus, in spite of the many difficulties that must be faced, there will be fewer but better steel companies. Unlike the grim outlook that sees the industry's problems worsening in the future, this alternative view contends that steel's worst days are past. As a result of decisions made during the current steel crisis to reduce capacity, staff, and overhead costs, as well as to install critical new facilities, the future looks much brighter. This opinion is shared by a number of steel companies, both integrated and non-integrated.

With a smaller steel capacity (most probably less than 130 million tons per year), less capital will be required to modernize that portion of capacity which requires updating. Since most of the recent shutdowns involved obsolete, high-cost equipment, the capacity that remains to be modernized is relatively smaller than it was at the beginning of 1981. For example, Bethlehem Steel shut down its Lackawanna plant which, to function economically, would have required a vast amount of money to replace most of its facilities, including a hot-strip mill of pre-World War II vintage. The replacement of this unit alone would have required $400 million. However, with the plant shut down, these funds can be used to update the hot-strip mill at Sparrows Point.

To assess the future of the steel companies in the United States, it is necessary to examine individual companies and plants as they are at present and will be in the next few years.

Present Composition of the Industry

The basic steel producing industry consists of a number of fully integrated plants with coke ovens, blast furnaces, steelmaking facilities, as well as rolling and finishing mills. In addition, there is a much larger number of electric furnace steel plants, that depend on scrap as a feed material. As of late 1983, the entire industry had a reliable capacity to produce approximately 140 million net tons of raw steel. It also had adequate facilities to roll and finish this potential production. The major portion of this steelmaking capacity (as well as finishing facilities) has been installed in recent years and consequently is competitive. However, as of late 1983, a significant portion, perhaps 30 percent, left much to be desired.

The integrated plants accounted for approximately two-thirds of the steel industry's capacity in 1983, while the electric furnaces, operating in both small and large plants, made up the other third. Steelmaking facilities in the integrated plants are overwhelmingly basic-oxygen furnaces with some small production contributed by the open hearth.

Because of the dominance in the industry of the integrated plants, as well as their vulnerability to foreign competition, this analysis will concentrate heavily on them. The electric furnace plants scattered throughout the United States have a degree of flexibility in relation to production and marketing that makes them less vulnerable.

An examination of the integrated segment of the industry reveals that many facilities are competitive and modern. These facilities include entire plants, such as the Burns Harbor Works of Bethlehem, as well as parts of plants where new, efficient equipment has been installed. Examples of these are the new seamless-pipe mill at the Fairfield, Alabama, Works of USS and the new rail mill at the Monessen Works of Wheeling-Pittsburgh. Much has been done in the last two decades to increase competitiveness, yet much remains to be done.

Unlike the Japanese, who built a new steel industry, U.S. companies, with the exception of two integrated plants, modernized and expanded existing facilities. To achieve this, some $40 billion were spent by steel companies in the United States in the last twenty years. Not all of this was put into steel and related operations. However, the major portion (at least $32 to $34 billion) was spent on steel plants and raw materials. Examples of the move to modernization are (1) the increase in continuous casting, which rose from 9.1 percent of steel production in 1975 to 31 percent in 1983, and

(2) the virtual elimination of the open hearth in favor of basic-oxygen and electric-furnace steelmaking equipment. Further, production practices and raw materials have been improved to the point where blast furnaces, which yielded 2500 tons a day in the mid-1970s, now produce in excess of 3500 tons.

In spite of the huge investments made during the last twenty years, there has not been enough capital available to make all of the integrated plants fully competitive. With the exception of Inland Steel, the major integrated producers have multi-plant operations and in the 1960s and early 1970s attempted to modernize all of these units. As a consequence, virtually all of them have some modern, competitive equipment, but at the same time they continue to operate some older, less competitive facilities since adequate funds were not available to make complete replacements. For example, one plant has fairly old blast furnaces but a modern oxygen-steelmaking shop and has just installed a new continuous caster. Its hot-strip mill is competitive, but its cold-reduction facilities leave something to be desired. Thus, in terms of modern equipment, the plant can be said to be out of balance. A number of other examples indicating similar conditions could be cited.

These situations are not so much the result of poor overall planning, but rather an attempt on the part of the companies to improve at all locations. The production improvements were made during the 1960s and early- to mid-1970s, when there was no thought of abandoning steel plants, but rather producers felt that all the plants could be fully modernized and rendered competitive. With the reduction in steel demand, the large tonnages of imports, and the depression that has befallen the steel industry, hindsight indicates that it perhaps would have been better to concentrate on fewer locations. However, when many of the investments were made, the present conditions were not foreseen.

Because of the shortage of capital, the drive directed to the improvement of steel plants would have fallen short of modernizing all of the facilities involved in the 140-million tons of steelmaking and finishing capacity in 1983. Consequently a number of companies deemed it necessary to abandon a portion of their facilities in order to improve and strengthen what remained.

U.S. Steel's December 27, 1983 announcement calling for the installation of some new facilities and the abandonment of others is directed precisely at this objective. The same can be said of similar announcements made by CF&I Steel (on the same day) and Republic (on January 18, 1984 when it abandoned its Buffalo facilities). The merger between Republic and Jones and Laughlin will bring about the closure of some additional facilities. However, the proposed acquisition of National Steel's facilities by U.S. Steel, now abandoned, would have resulted in minimal closures.

A survey of the integrated companies and their plants in the United States will demonstrate what has been done and indicate what remains to be done for the companies to achieve a better, fully competitive status. As of the end of 1983, there were fourteen companies operating thirty-two integrated plants. The number of companies will be reduced by the Republic/J&L merger. Further, in 1984 and 1985, a number of plants will be closed or scaled down in size and thus removed from the integrated category.

Since the integrated plants will be the subject of restructuring much more than the electric furnace operations, this book will deal predominately with integrated plants operating coke ovens, blast furnaces, and basic oxygen steelmaking units. However, a brief analysis of the growth of the electric furnace in both the integrated companies and minimills is given.

The thirty-two integrated steel plants (as of December 1983), spread over fourteen companies, had the following distribution:

1. *U.S. Steel Corporation*
   Gary South Works at South Chicago and Gary, Indiana
   Lorain Works at Lorain, Ohio
   Fairless Works at Fairless Hills, Pennsylvania
   Fairfield Works at Fairfield, Alabama
   Pittsburgh Works in the Pittsburgh, Pennsylvania, area
   Geneva Works at Geneva, Utah
2. *Bethlehem Steel Corporation*
   Bethlehem Works at Bethlehem, Pennsylvania
   Sparrows Point Works at Baltimore, Maryland
   Burns Harbor Works near Chicago, Illinois
3. *Inland Steel Company*
   Indiana Harbor at East Chicago, Indiana
4. *Armco, Inc.*
   Middletown Works at Middletown, Ohio
   Ashland Works at Ashland, Kentucky
5. *National Steel Corporation*
   Great Lakes Division at Trenton, Michigan
   Granite City Works at Granite City, Illinois
   Weirton Works at Weirton, West Virginia
6. *Jones and Loughlin Steel Corporation*
   Aliquippa Works at Aliquippa, Pennsylvania
   Cleveland Works at Cleveland, Ohio
   Indiana Harbor Works at East Chicago, Indiana
7. *Republic Steel Corporation*
   Cleveland Works At Cleveland, Ohio
   Warren Works at Warren, Ohio
   Chicago Works at Chicago, Illinois

Gadsden Works at Gadsden, Alabama
Buffalo Works at Buffalo, New York
Youngstown Works at Youngstown, Ohio

8. *Wheeling-Pittsburgh Steel Corporation*
   Monessen Works at Monessen, Pennsylvania
   Steubenville Works at Steubenville, Ohio
9. *McLouth Steel Products Corporation*
   Trenton Works at Trenton, Michigan
10. *Rouge Steel Company*
    Works at Dearborn, Michigan
11. *Sharon Steel Corporation*
    Victor Posner Works at Sharon, Pennsylvania
12. *CF&I Steel Corporation*
    Pueblo Works at Pueblo, Colorado
13. *Interlake, Inc.*
    Blast furnaces at Chicago, Illinois, and steel operations at Riverdale, Illinois
14. *Lone Star Steel Company*
    Lone Star Works at Lone Star, Texas

United States Steel

Since its founding in 1901, U.S. Steel (USS) has made a number of acquisitions and undergone several changes and reorganizations. Many of these were significant, particularly when subsidiary companies were combined. However, none has affected the organizational structure as profoundly and dramatically as have those made in 1982 to 1984.

In early 1982, USS made a major acquisition when it bought Marathon Oil Company, a large, integrated, domestic company involved in exploration, production, refining, and distribution of petroleum products. Marathon also has substantial international operations. The price was $5.9 billion, of which $3 billion were borrowed and $2.8 billion worth of notes were issued. In terms of sales volume, Marathon was the largest acquisition made in the history of the corporation. In 1982, its sales of $9.6 billion constituted one-half of the entire USS revenue for that year. This comparison must be qualified to some extent: 1982 was a very poor year for steel shipments as they hit a forty-year low of ten million tons. The acquisition was aimed at improving the profit position of the corporation and protecting it against the downside cyclical fluctuations in the steel industry. This was achieved in 1983, when Marathon's operating income was $1.1 billion on sales of $9.3 billion.

Although the Marathon acquisition greatly accentuated USS's diversification, it did not represent a new trend as USS has been diversified to a greater or lesser extent for many decades. In 1950, the USS Annual Report listed some thirty-three principal subsidiaries. These included a number of steelmaking units as well as those engaged in the distribution and fabrication of steel. There were also railroads and other types of transportation as well as one of the nation's largest cement companies. Some of these subsidiaries were merged in the intervening years and others were acquired so that by 1982 the annual report listed twenty-five divisions and subsidiaries.

Most of the subsidiaries were closely connected with the production of steel. For example, the railroads (of which there were five) serviced the steel operations, although to some extent they acted as common carriers. The same can be said of several other subsidiaries engaged in water transportation.

Nonsteel activities (before the acquisition of Marathon Oil) were principally in chemicals, cement, and real estate. The Universal Atlas Cement Company was a part of USS from its early stage. Chemicals were a more recent operation for, although the corporation marketed some of the by-products from its coke ovens, a major step was taken in 1964 when USS purchased substantial segments of the Pittsburgh Chemical Company. This division produces basic materials for the plastic industry and uses some raw materials that are by-products of the coke ovens. In 1969, the chemical activities were reorganized to form two units: USS Agrichemicals and USS Chemicals. These operations were expanded at that time with the inauguration of five new chemical plants. Alside, a company engaged in housing construction, was acquired in 1968.

In 1982, USS consisted of the following divisions:

1. Steel, including not only steel products but domestic ore and coal operations.
2. Oil and gas, including Marathon and its subsidiaries.
3. Chemicals, including the production and marketing of coal chemicals, petrochemicals, plastic resins, and agricultural chemicals.
4. Resource development, including commercial development of mineral and energy resources in excess of USS requirements, as well as exploration for new mineral and energy resources.
5. Manufacturing and other, including steel service centers, real estate development, and the manufacture of products for residential construction.
6. Domestic transportation and utility subsidiaries, including commercial carrier railroads, domestic barge lines, gas utilities, and a dock company.[1]

During the past three years, USS decided to divest itself of a number of activities so that it can concentrate its efforts in fewer categories. This program resulted in the sale of the following:

1. Universal Atlas Cement Division.
2. A considerable amount of real estate, including the sixty-two-story office building in Pittsburgh that houses USS corporate offices.
3. A 50 percent interest in Navios, an ocean-shipping line.
4. USS Products Division, with its five plants engaged in manufacturing pails and drums.
5. Extensive coal reserves above the needs of USS.
6. Alside, a manufacturer of housing materials.
7. The electric-cable division.
8. The tire-cord division.

Despite these divestitures, as of 1984, USS stands as a highly diversified company.

In terms of its steel operations, there have been a number of drastic changes in the last few years. In 1979, the corporation wrote off steel facilities considered obsolete or (at best) marginal. The major portion abandoned was the 1.7-million-ton-capacity integrated plant (except for coke ovens) at Youngstown, Ohio. In addition, facilities such as the plate mill at Fairfield, Alabama, the 80-inch hot-strip mill at Gary, Indiana, and the rod mill at Pittsburg, California, were written off. From 1979 through 1984 the emphasis has been on improving efficiency, cutting costs, and reducing rather than increasing the size of the steel division.

It became evident in 1983 that a number of additional facilities, some of which had been temporarily closed down for business reasons, were unprofitable and obsolete. The amount of money that would have been required to replace or modernize these units was enormous and projections of the steel market for years ahead did not indicate sufficient strength to warrant this investment. The one exception was the construction of the new seamless-pipe mill and accompanying continuous caster at Fairfield, Alabama. This was deemed necessary if the corporation was to maintain a competitive position in the oil-country goods market.

After an analysis of the steel sector by the corporation, sweeping changes affecting virtually all of these plants were announced on December 27, 1983. These changes included the installation of two new continuous casters, the reactivation of the Fairfield, Alabama, plant, and the closure of a large number of facilities, including two basic iron and steel operations. The move was "designed to insure that its steelmaking sector will continue as a major force in world steel markets for the balance of the century and beyond."[2] It was announced that the two continuous casters were to be located at Gary, Indiana, and Fairfield, Alabama.

The closures included a number of plants and parts of plants, among which were the blast furnaces and oxygen-steelmaking shops of South Works in Chicago and the National-Duquesne Works near Pittsburgh. These, along with some other relatively minor abandonments, reduced the steelmaking capacity of USS from 31.0 million to 26.2 million tons, a loss of approximately 15 percent. Other plants that were completely shut down included the Johnstown Works at Johnstown, Pennsylvania (a relatively small electric furnace plant with two steel foundries) and the Cayuhoga Works in Cleveland (a rod mill producing rods and facilities for wire and cold-rolled strip). Other major facilities marked for closure included the rail mill at Gary, Indiana, and rod mills at South Chicago and Fairless.

Some 15,000 employees were terminated; over 10,000 of these were already on layoff, however. This number of layoffs was neutralized to some extent with the resumption of operations at Fairfield, bringing with it some 1600 jobs; 800 more were created by the installation of the new pipe mill at Fairfield.

The philosophy behind the announcement on December 27, 1983, is summed up in the 1982 10-K Report which states:

> The Corporation seeks to allocate its capital resources selectively to the businesses that best provide opportunities. More than half of the capital budget is currently directed to oil and gas segment projects. Capital spending in steel, which increased in 1982 over 1981 and was comparable to 1980, will continue to be directed toward cost reduction and improvements in productivity, energy efficiency, product quality, and customer service, in addition to environmental and other legally mandated expenditures.[3]

A list of the major steel projects that were brought into operation in 1982 and 1983, as well as those projected for subsequent years, bears out this statement. In 1982, a large new coke-oven battery was put in operation at Clairton to replace older, less efficient units. There were also improvements to the existing continuous caster at Gary, as well as to the emission controls on coke-oven batteries at several plants. In 1983, the main installations included a continuous caster at Lorain, Ohio, which would improve quality and yield, and the completion of a new seamless-pipe mill at Fairfield, which would allow the corporation to meet competition in that product area. Other investments consisted of improvements to operating units.

Projects to be completed beyond 1983 consisted of finishing the renovation of a blast furnace that was approximately 80 percent complete; improvements to a sinter plant and specialty-tube mills at Gary; and the first continuous caster at Fairfield, which would accompany the pipe mill. Two other continuous casters for slabs, one at Fairfield and the other at Gary, were also included. A review of these investments indicates no steel expansion but rather modernization and improvements of facilities to permit higher quality and lower cost steel production. The changes, which resulted in the reduction of plant size, indicate that in the future USS will have smaller (and hopefully more efficient) capacity to produce steel.

The move to reducing the size of steelmaking capacity was countered by a dramatic announcement on February 1, 1984, that USS would acquire the steel facilities of National Intergroup. These facilities included two fully integrated plants (one near Detroit and the other near St. Louis) as well as a modern finishing facility near Chicago; total capacity would be approximately 5.5 million tons. Thus USS would have been able to bring its capacity up to the 31.0-million-ton figure it had before the December closings were announced. Unfortunately the Justice Department opposed the acquisition and the plan was dropped.

A plant-by-plant analysis indicates the extent of USS's restructuring, as well as its future plans for the steel business.

Fairfield Works

The Fairfield plant, located near Birmingham, Alabama, was closed down in mid-1982 when the union and management failed to reach an agreement on the number of employees needed to run the plant efficiently. On December 27, 1983, USS announced that it had come to an agreement with the union and would reopen the plant early in 1984. During the shutdown period, construction continued on a modern seamless-pipe mill and a companion continuous caster; the total investment was about $700 million. The pipe mill was put in operation on a limited basis in late 1983, and the continuous casting unit is scheduled to function in 1984.

At the time of the shutdown, the Fairfield Works had two coke-oven batteries, one of which was relatively new (installed in 1978 with a capacity to produce 900,000 tons a year). Unfortunately, it had suffered considerable damage before being closed and will require extensive repairs. There were also two blast furnaces, one of which was a 5,400-ton-a-day unit installed in 1978. This furnace operated very efficiently from its start-up to the close down four years later. There was also a modern steelmaking shop (installed in 1978) with bottom-blown oxygen converters capable of producing somewhat in excess of 3.0 million tons of steel a year.

The finishing facilities consisted of a hot-strip mill and cold-reduction facilities, along with facilities for the production of hot and cold rolled sheets, tinplate, and galvanized sheets.

Successful negotiations with the union on manning the plant have brought the entire facility back into operation in early 1984. This reactivation was a necessity in order for the pipe mill to operate efficiently as part of an integrated plant. The pipe mill has a capacity to produce 600,000 tons of product; this would require some 700,000 tons of steel. However, with this amount of production, the steelmaking shop, with a capacity of 3.0 million tons, would be operating at a totally-uneconomical less than 25 percent of capability. The reactivation of the entire plant will allow the steelmaking facilities to operate at or near their full potential.

Fairfield, with its new seamless-pipe mill fed by a continuous caster and its reactivated sheet production should be competitive. This situation should further improve with the announced continuous caster for slabs. However, it will take the better part of two years before this unit is in place and operating.

In addition, extensive repairs will have to be made to the large coke-oven battery; this will take at least eighteen months, during which time coke will be supplied from the Clairton Works near Pittsburgh. Some modernization will be necessary for part of the hot-strip mill as well as other finishing facilities. Fairfield has two cold-reduction mills, one of which is a six-stand mill installed in the 1960s. This mill is one of three in the United States and is particularly adaptable for the production of light-gauge tinplate. The improvements made over the next two years, once completed, will permit the plant to operate with a high degree of efficiency. By the mid- to late 1980s, Fairfield should be an efficient, profitable unit, producing 3.0 million tons of raw steel and capable of shipping at least 2.5 million tons annually. Fairfield will continue to use Canadian and Venezuelan pellets high in iron content, which replaced local ores in the early 1960s.

Some observers of the steel industry indicate that the plant operates at a disadvantage, since it is not located on the seacoast. This may be true to some extent; however, the plant historically was placed near Birmingham within five miles of its iron and coal supply. The iron ore was relatively low grade, but its proximity to the plant compensated in great part for the low quality. It is still close to the source of coal supply, and ore from Venezuela will be shipped to the plant from Mobile. It would be financially impossible to relocate the Fairfield facilities. They can be operated efficiently in their present location, particularly with the new slab caster.

Lorain

The steel plant at Lorain, Ohio, is located on Lake Erie and has the advantage of receiving its ore supply by water. The plant is principally a pipe producer with several mills including three for seamless pipe. Pipe production is supplemented by the output of two modern bar mills. Lorain has also been tied intimately to the Cayuhoga Works in Cleveland to which it shipped steel billets to be rolled into rods, wire, and narrow strip. Unfortunately the Cayuhoga plant, with a capacity of 500,000 tons, was closed down permanently in April 1984, thus curtailing Lorain's raw-steel output.

In October and November of 1983, the coke-oven batteries at Lorain were shut down permanently. Henceforth Lorain will receive coke for its blast furnaces from the Clairton Works in the Pittsburgh district.

Presently, there are four blast furnaces standing at the plant. Numbers 1 and 2 are relatively small and will be phased out in the next few years. Number 3 will undergo a complete renovation in 1985, thereby increasing its capacity to 3,600 tons a day and improving the efficiency of its operation. When this is completed, number 4 will most probably be renovated in a similar manner and the plant will then operate on two completely rebuilt, modern blast furnaces.

The basic-oxygen steelmaking facilities are modern and have an annual capacity of 2.8 million tons. Further, the plant has just installed a continous-casting machine that will produce both rounds for seamless tubes and billets for the bar mills.

Lorain figures predominantly in the future plans of USS; with its blast furnaces renovated and the continous caster in operation Lorain should be a competitive facility producing seamless pipe and tubes as well as bars. This plant, combined with the new seamless mill at Fairfield, Alabama, will cover the complete range of seamless tube products.

Geneva

The mill at Geneva, Utah, near Salt Lake City, was built as a government plant during World War II in order to provide plates for shipyards on the West Coast. In addition to the plate mill, it had a structural mill, which recently has been abandoned. The plant was located in Utah rather than on the Pacific Coast in order to be out of range of possible Japanese bombing attacks against the West Coast. In terms of raw materials, the location was quite good since there was ore in southern Utah and coal was available within a reasonable distance.

Currently, the mill has four coke-oven batteries, three blast furnaces, an open-hearth shop with ten 340-ton furnaces, and a continuous plate and wide hot-strip mill. The original rolling facility was a 132-inch plate mill. However, at the close of World War II, when USS acquired the entire plant, it rebuilt the plate mill to roll sheets as well as plates. These two items constitute the current product mix.

The coke-oven batteries, as well as the blast furnaces, have functioned for forty years. At present, only two of the blast furnaces are operating; the third is held in reserve temporarily. One coke-oven battery needs attention; however, the other three are in good condition and adequate to supply the needs of the blast furnaces.

In the early 1960s, USS built a pellet plant at Atlantic City, Wyoming, intended to furnish iron ore pellets for the Geneva blast furnaces. This plant functioned up to 1983; then it was closed permanently. Currently, ore is being supplied from the USS Minntac pellet plant in Minnesota whose pellets are superior to those formerly received from Atlantic City, thus improving blast furnace operation to a point where part of the additional freight from Minntac can be neutralized.

Steelmaking at Geneva will continue with the open-hearth process. However, some modifications which will require EPA approval are necessary. Currently, no major investment is needed to keep the plant's 2.6 million tons of steelmaking in operation.

The Geneva Works has been intimately tied to the USS plant at Pittsburg, California, to which it ships most of its production of hot-rolled sheets. The Pittsburg, California, plant processes these through its cold-reduction facilities, after which much of the product is galvanized or tinplated.

This plant is USS's major facility on the Pacific Coast from which it serves that market. Consequently, the Geneva plant must continue its operations if USS is to service the West Coast sheet and tinplate market. Without hot bands from Geneva, other sources of this material would have to be found, which would be difficult to achieve on a permanent basis.

USS's 1982 Annual Report announced a new alignment of steel plants. In respect to the West, it states:

> In the West, the Geneva (Provo, Utah) and Pittsburg (Calif.) plants were combined. This move will improve the planning and coordination of production at both plants in serving customers throughout the western states.[4]

Some observers of the industry questioned the continued existence of Geneva as a source of hot-rolled bands for the Pittsburg Works. However, it is difficult to see how USS can obtain a permanent supply of hot-rolled bands from another source on an economical basis. If these were to come from somewhere in the United States, they would have to be shipped from the East or Midwest to California. The Midwest would seem to be out of the question because of the rail-freight rate; it is difficult to see how the bands could come from the East Coast, since the Fairless Works would not have the capacity to supply the full requirement, and freight would be costly.

In terms of a foreign supply, it is hard to see how a permanent source could be found for the amount of tonnage necessary. Undoubtedly, in times of slack world demand, hot-rolled bands could be obtained. However, in any surge in world steel activity, it would be very difficult to obtain them at reasonable prices. One possibility, of course, would be a major investment in a foreign steel company. For the near future, Geneva will continue to operate; however, some decisions must be made in respect to its basic facilities sometime in the next five years.

Fairless Works

The Fairless Works, located in eastern Pennsylvania (near Trenton, New Jersey), was erected between 1950 and 1952 in order to give USS better access to the East Coast market. The plant is one of two fully integrated greenfield-site plants built in the United States since the close of World War II. When completed in 1952, it employed the latest technology of that day and was considered the most modern in the United States. It has a now-inadequate sinter plant and two batteries of coke ovens that can produce a total of 800,000 tons of coke per year, not quite sufficient to satisfy the plant's requirements at full operation. The coke ovens were shut down in the first quarter of 1984, so that all of the coke for Fairless is brought in from Clairton. This is a temporary condition; however, in 1986 or 1987 the Fairless coke ovens should undergo major rebuilds or replacements. Instead, it is quite possible that a decision may be made to abandon them and continue to supply coke from Clairton.

There are three blast furnaces, two of which were built with the original facility, while the third was added in 1957. All three have been relined in the late 1970s and early 1980s. When first installed, each furnace was capable of producing approximately 550,000 tons annually. Through the years, with the improvement in furnace burden and practice, the capacity has been expanded, so that in 1982 they were capable of producing 1.0 million tons each per year.

The steelmaking facilities consist of an open-hearth shop with nine furnaces, each with a capacity of 410 tons per heat, making for an annual capacity of approximately 3.5 million tons. In addition to the open hearth, there are two electric furnaces, each with a capacity of 200 tons per heat. The open-hearth shop was installed at the time of the original construction, which was started in 1950 when the open-hearth process was the predominant steel producer and the basic-oxygen converter had not yet been developed. Electric furnaces and a continous caster were added to the plant's facilities in 1972 and 1973.

The principal finishing facility is an 80-inch wide hot-strip mill which processes slabs into hot-rolled coils. Some of these coils are sold as such, while most of the output is further processed into cold-rolled sheets, tinplate, and galvanized sheets. Fairless also has two pipe mills for continuous buttweld pipe, as well as a rod mill that was installed in 1969 with a 500,000-ton annual capacity.

The plant became a center of controversy when it was announced that USS and British Steel Corporation (BSC) had entered into conversations for a possible arrangement whereby the iron and steelmaking facilities at Fairless would be shut down and 3.0 million tons of slabs provided from the Ravenscraig plant of BSC, located in Scotland, which, in turn, would close down its finishing facilities. This arrangement seemed logical since the Ravenscraig plant of BSC was a candidate for closure and the iron and steelmaking facilities at Fairless have become relatively obsolete and would have to be replaced at a very high cost. Thus an integrated plant of a sort would have been created with the melting end in Scotland and the finishing end in eastern Pennsylvania.

During the ensuing controversy, the USS management stated that it had no intention of modernizing the iron and steel facilities at Fairless. Such a project would have involved a new sinter plant, considerable adjustments to the blast furnaces and coke ovens, the installation of a basic steelmaking shop, and the construction of a continous caster for slabs. The cost of these installations would have been approximately $1.0 billion, and corporation management did not wish to commit that much capital to one plant when they felt that it could be used to greater advantage on other steel facilities.

The arrangement with British Steel called for a considerable investment of some $600 million to be made by the British, a part of which would be used to improve the rolling and finishing facilities at Fairless. It was stated that failing this arrangement the iron and steelmaking equipment at Fairless would continue to operate until it reached the end of its useful life.

On December 27, 1983, USS announced that the negotiations with British Steel had terminated since a satisfactory agreement on the cost of slabs to be provided to Fairless could not be reached. However, USS said it would continue to seek slabs for the Fairless Works from other sources. This did not necessarily mean that the entire requirement of Fairless would be supplied from outside; more likely it would only be part of it.

The December 27 announcement also said that the rod mill and wire facilities at Fairless (as well as one electrolytic tinning line) would be permanently closed down. The rest of the plant's short-run future (that is, the next two to three years) seems to be assured. However, in the long run, the plant's future will depend to some extent on an outside supply of semi-finished steel to furnish part of its requirements. This could be obtained from either domestic or foreign sources. At present, it seems that the Fairless finishing facilities, which can absorb 3.0 million tons of semifinished steel, will continue in operation on a combination of in-plant slabs and slabs from other locations. How much of each will be determined to a great extent by the price of scrap. If the price is relatively low, the open hearth will be able to produce at an economical cost and satisfy the full requirement; if the scrap price is high, the steelmaking costs will increase accordingly and production may be cut back in favor of outside slabs.

As the years progress, there will probably be fewer in-plant slabs provided and more from the outside. Thus, by about 1990, since there will be little or no investment in iron and steelmaking facilities, output of steel may decline, and the finishing facilities will function with a greater dependence on outside steel supply.

Pittsburgh District

In 1981, there were twelve operable blast furnaces in the Pittsburgh district. These were located in groups of four at three steelmaking plants—namely, Homestead, Edgar Thomson, and National-Duquesne Works.

At Homestead, in addition to four blast furnaces, there was an open-hearth shop, built during World War II, with eleven furnaces each capable of 320 tons per heat. There were also blooming and slabbing mills as well as two plate and two structural mills. In addition, Homestead had a special division for making forgings with steel obtained from the electric furnaces at National-Duquesne Works. Since 1981, the four blast furnaces and the open-hearth shop, with over 3.0 million tons of capacity, have been closed down so that the plant is now a finishing mill, rolling steel (obtained from Edgar Thomson) into plates and structurals.

In 1981, the Edgar Thomson plant listed four operable blast furnaces; a basic-oxygen steelmaking shop installed in 1972; and a slabbing mill. At that time, the plant was tied to the nearby Irvin Works, which was a complete sheet finishing facility built in 1938. It has an 80-inch hot-strip mill, several cold-reduction mills, as well as tinning and galvanizing equipment. In the 1980 Iron and Steel Directory published by the American Iron and Steel Institute, the two plants are listed together as Edgar Thomson–Irvin Works. The blast-furnace segment of the Edgar Thomson Works was reduced to three in 1981, when the smallest furnace was closed permanently. The smallest remaining unit underwent a complete renovation and rebuilding, which as of the end of 1983 was about 80 percent finished. The investment in the furnace thus far is about $60 million, and when complete, it will be a thoroughly modern, efficient although relatively small (i.e., 25 feet in hearth diameter) unit. The other two blast furnaces at Edgar Thomson will continue to operate. However, the slabbing mill has been mothballed and the plant now ships ingots to Homestead where they are rolled into slabs for Homestead and Irvin.

The National-Duquesne Works listed four blast furnaces in 1980; one, Number 6, was the largest in the Pittsburgh district. In addition, the plant had a basic-oxygen steelmaking shop that was installed in 1963 and was the first to be built by USS. The output of the plant consisted principally of seamless pipe and tubing, as well as a variety of bars. In 1981, two of the blast furnaces were abandoned and a third was taken out of operation, leaving Duquesne Number 6 as the only functioning unit.

Another major plant in the Pittsburgh district is the Clairton Works, which is confined almost exclusively to the production of coke. It is one of the largest coke plants in the Western world with a capacity in excess of 5.0 million tons. The plant is supplied with coal to be turned into coke from the USS mines in the Appalachian coal region.

The plants in the Pittsburgh area were put under one management in 1982. The annual report for that year states:

> In the Pittsburgh area, a new Mon Valley Works was created, consisting of four plant units: a unified Edgar Thomson-Duquesne plant with blast furnaces, basic oxygen and electric furnaces for iron and steel production, plus a bar mill and an iron foundry; the Homestead plant with plate and structural rolling facilities and related operations; the National plant with tubular operations; and the Clairton plant with coke and chemicals operations.[5]

On December 27, 1983, this was changed somewhat when USS made its announcement restructuring its steel facilities. The Pittsburgh area was materially affected. The blast furnaces and steelmaking equipment at National-Duquesne Works were abandoned with the blooming mill and Number 1 seamless pipe mill. Number 2 seamless mill will remain operable. Henceforth, all the iron and steel in the Pittsburgh district will be made at the Edgar Thomson plant whose basic-oxygen steelmaking shop is capable of approximately 3.0 million tons annually. The Homestead Works will continue as a rolling mill for plates and structurals and Clairton will continue to produce coke for the Edgar Thomson blast furnaces, as well as those located at the Lorain Works near Cleveland. It will also supply some coke to the Fairless Works in Eastern Pennsylvania and at least through the end of 1985 to the Fairfield Works near Birmingham, Alabama.

The electric furnaces at Duquesne will continue to furnish special steel for the Homestead plant while the Irvin hot-strip mill will be provided with slabs from Homestead and occasionally will receive hot-rolled bands from either Fairless or Gary.

The Pittsburgh district has been materially reduced in capacity with the loss of approximately 2.7 million tons of steelmaking capacity at the National-Duquesne Works. To some extent, it will be dependent on plants outside of the area for some of the steel to operate the Irvin Works, which will continue to produce hot and cold-rolled sheets as well as tinplate. The restructuring of the Pittsburgh district, in effect, results in one integrated steel plant, with about 3.0 million tons annual capacity, composed of the remaining units, all of which are located close to each other. Employment will be reduced and opportunities for future employment will be limited. There are no major investments planned for the next few years. It is conceivable that within four or five years a continous casting machine could be installed at Edgar Thomson. However, although this has been discussed, no decision has been made.

Chicago Area

The Chicago area has two principal plants: the Gary Works in Gary, Indiana and the South Works in South Chicago. They were merged under one management in 1982, and at that time, the Irvin Works in the Pittsburgh area were included as part of this unit. The reason for this was the closure (on what proved to be a temporary basis) of the hot-strip mill at Irvin. Hot-rolled bands from the Gary strip mill were supplied to the cold-reduction facilities at the Irvin Works. As stated in the 1982 Annual Report: "The South (Chicago) and Irvin (Dravosburg, Pa) plants were merged into an expanded Gary (Ind.) Works. All three plants have considerable production compatibility and have had various operating interrelationships for many years."[6]

Gary

The Gary steel works was at one time the largest steel plant in the United States and, for that matter, in the world. It had twelve blast furnaces and was capable, with its open-hearth shops, of producing in excess of 8.0 million tons of steel per year. Finished products included sheets, tinplate, plates, bars, rails, and railroad car wheels. At the present time, the plant has been streamlined with seven of its original furnaces shut down and one more scheduled to be shut down. Currently, including the new Number 13 furnace, there are six furnaces operable. Number 13, the largest in the USS was built in 1977. However, due to some problems, the furnace was virtually rebuilt in 1980 and since then has functioned very well. In April 1984, only five furnaces will remain operable since Number 10 is scheduled to be abandoned.

Gary is fully integrated, with coke ovens having about 3.0 million tons of capacity to feed its blast furnaces. Some major rebuilds in the blast furnace sector will be forthcoming in the next two to three years. This program will involve the four furnaces (other than Number 13) and will cost at least $350 to $400 million. The present ironmaking capacity at Gary is about 5.5 million tons. Steel production comes from two basic-oxygen shops installed in the early 1970s. One, with three 215-ton conventional top-blown vessels, has an annual capacity of 4.0 million tons; another, with three 215-ton Q-Bop or bottom-blown oxygen converters, has an annual capacity of 3.5 million tons. Currently, there is one continuous casting unit, which processes about 40 percent of the output from the conventional BOF shop, and a second unit that has been announced will allow the plant to cast 50 percent of its total steel output.

In terms of finishing facilities, the Gary plant will continue to operate a modern 84-inch-wide hot-strip mill and a number of cold-reduction facilities for sheets. The plant will also produce tinplate and galvanized sheets as well as plates and bars. According to the December 1983 announcement, Gary will lose one of its oldest operating units, the rail mill. This was scheduled to be replaced by a new mill at the South Works; however, a decision has been made not to build that mill. In addition, Gary will lose the aforementioned blast furnace, three bar mills, and billet and blooming mills.

South Works

During the 1950s, South Works had eleven operable blast furnaces with a total ironmaking capacity in excess of 4.0 million tons. It had three open-hearth shops, as well as Bessemer converters and relatively small electric-arc furnaces. Total steelmaking capacity was approximately 5.5 million tons.

Since that time, the plant has been modernized with the installation of three 200-ton basic-oxygen vessels with a rated capacity of 4.0 million tons annually. The number of blast furnaces has been gradually reduced to two, one of which (Number 8) has been completely renovated in the 1980s with the capability to produce 4,000 tons of iron per day. Number 12 (built in 1948) has had a number of relinings, the last of which was in 1981. Since mid-1982, both blast furnaces and the basic-oxygen shop have been idle due to poor business conditions. Consequently, since that time steel at South Works has been produced in electric furnaces, of which there are three with a total capacity somewhat less than 1.0 million tons. In terms of finishing facilities in 1983, there were two plate mills, two structural mills, and a rod mill.

With the announcement in December 1983, South Works was dramatically affected. The blast furnaces and BOF vessels, which had been inoperative for a year and a half, have been closed down permanently as were the rod mill, both plate mills, and one structural mill. The remaining facilities consist of electric furnaces, a blooming mill, and a wide-flange structural beam mill. Thus, South Works ceases to be an integrated steel plant. The rail mill that was conditionally projected for the plant has been cancelled. The future capacity of the plant will be less than 1.0 million tons of raw steel, and its output will be limited to large structural sections.

Other Steel-Producing Units

In addition to the integrated steel works, the only other major steel-producing facility is the electric-furnace operation at Baytown, Texas. This has four electric furnaces (two rated at 200 tons installed in 1970 and 1971), along with a slab caster, and a 160-inch plate mill, capable of producing 1.3 million tons of plate annually. In 1977, two additional 220-ton furnaces and two slab casters were added. These provide enough steel to operate the plate mill at its capacity. Another major finishing facility at the plant is a pipe mill for welded large diameter pipe up to 48 inches.

Baytown will continue to function as an important part of USS operations. The plate mill is now the only large facility of its kind in the Southwest.

Another electric-furnace installation at Johnstown, Pennsylvania, which provided steel for castings, was shut down in April 1984.

Restructuring of United States Steel

USS has reduced its steel capacity and restructured its operations to become more competitive. The reduction of capacity to 26.2 million tons, much of which is of recent vintage, should make it possible for the corporation to reduce its break-even point to 60 percent of capacity. The restructuring, as indicated, involved the closure of a number of facilities, the installation of continuous casting, and closer coordination among the plants in the same area. Between 1981 and 1984, USS closed down twelve blast furnaces and by 1986 plans to close two more. It also made drastic changes in the Pittsburgh and Chicago districts. However, the new structure still requires some implementation, and it seems that additional small facilities may be eliminated.

The Chicago district, with 7.0 to 8.0 million tons of steel at Gary and somewhat less than 1.0 million at South, will continue to be the largest unit in the corporation. All other areas, including Pittsburgh, Fairfield, Fairless, Geneva, and Lorain, have capacities ranging from 2.6 to 3.5 million tons.

The decision to close the rod mills at South Chicago, Fairless, and Cayuhoga effectively puts the corporation out of the rod business. The only remaining unit is a relatively small mill for the production of special quality rods at Joliet, Illinois. Competition from minimills has made rods an unprofitable business; dropping them will eliminate a loss item. Remaining rod producers include Raritan Steel in New Jersey, Georgetown Steel in South Carolina, and Georgetown Texas. The last of these was sold to Cargill. Before the sale, the Texas plant sold rods at a loss for less than $300 a ton, which made it unprofitable for larger companies to compete.

USS will also cease to produce rails with the closure of the Gary rail mill in April 1984. The announcement of the new rail mill at South Works in Chicago was cancelled when it was stated that adequate concessions, deemed necessary for the mill to function economically, were not made. The only possibility for reentering the rail market is through the purchase of a rail mill. Here there seems to be but one option: the mill of CF&I Steel at Pueblo, Colorado. This is one of three rail mills that will remain after the closure of Gary. The other two are at Bethlehem's plant at Steelton, Pennsylvania, and Wheeling-Pittsburgh's plant at Monessen, Pennsylvania. If the corporation buys CF&I's rail mill, it will not increase the number of producers. As indicated elsewhere, the rail market has been weak for ten years, averaging a little over 1.0 million tons per year. In 1982, it dropped to half of that figure and, although it might return to a million tons, it is difficult to see how it will be much more than that figure in the years ahead. Consequently the three mills have adequate capacity to meet the railroads needs.

The two continous casters announced for Fairfield and Gary will add a significant measure of efficiency to operations at those plants. The construction of a modern world-class seamless pipe mill with accompanying continous caster at Fairfield, Alabama, coupled with the reopening of that plant, should improve the picture at that location—particularly when the new caster for slabs is in operation. There is still, however, a significant investment required in the finishing facilities at that plant to make them fully competitive. Other investments that will be required include renovation to the blast furnaces at Lorain and Gary. Further, a decision concerning the Geneva plant will have to be made within the next three to four years. For the immediate future, the plant will run as it has for the past forty years. The elimination of so many blast furnaces throughout the corporation and the improvement of the remaining units should reduce iron costs significantly, particularly with higher coke stability which lowers the coke rate.

One question mark in terms of the future is the Fairless Works. Its open hearth will continue to provide steel for the years immediately ahead. However, it is quite conceivable that the steel supply may be supplemented with slabs from outside the plant, produced either domestically or by foreign mills. This will probably not be necessary until at least 1985 or 1986. However, should the price of scrap increase significantly, rendering open hearth operations less economical, a portion of the slabs may well be brought into the plant before that time.

In terms of raw materials, USS has more than an adequate supply with several billion tons of ore reserves located in the United States and Canada. Much of this ore is low grade, which must be concentrated in pellet plants; the largest of which is Minntac in Minnesota, with a capacity to produce 21.0 million tons of pellets. Further, the corporation has interests in Canadian operations and a commitment to the Venezuelans to buy ore for Fairfield and Fairless.

In respect to coal, USS has over 2.0 billion tons of reserves located principally in Alabama, Kentucky, Virginia, West Virginia, and Pennsylvania.

In 1983, USS embarked on a program that by 1984 or 1985 should permit it to break even on shipments of 1.0 million tons a month. When business activity in steel exceeds that, the corporation should return to profitability in its steelmaking segment. The program involves a number of items, including reducing labor costs per ton, improving productivity as well as yield, reducing the energy consumed, upgrading facilities, and, where necessary, eliminating obsolete plants and equipment.

Labor cost per ton can be reduced in two ways: (1) a cut in hourly wages which was achieved in early 1983, and (2) an improvement in productivity which will result in fewer man-hours per ton and thus reduce labor cost per ton. In an attempt to cut labor costs, the adopted program affects management as well as unionized employees and aims to reduce administrative costs.

The acquisition of Marathon and the attempted acquisition of National Steel's facilities highlight the corporate changes at USS.

Acquisition of Marathon Oil

The acquisition of Marathon Oil by USS was the single most significant change in terms of diversification and restructuring that has taken place since the foundation of the corporation in 1901.

Marathon is a completely integrated operation involved in the exploration, production, refining, and distribution of oil and petroleum products. It also has significant international interests. The company has large oil and gas reserves, including a 49 percent ownership in the extensive Yates Field in Texas. This field is second largest in production in the United States, with current allowables of 125,000 barrels per day. Marathon is also very substantially involved in the North Sea Brae field which began production in 1983 and will add some 40,000 barrels of oil per day to the company's total production.

Marathon has four refineries in the United States with an aggregate capacity of almost 600,000 barrels of crude oil per day, ranking it ninth in refining in the country. Marathon markets petroleum products in the Midwest: Illinois, Indiana, Kentucky, Ohio, Michigan, and Wisconsin.

The price of $5.9 billion was by far the largest sum ever paid for an acquisition by USS. To consummate the transaction, it was necessary for USS to borrow $3.0 billion and issue $2.8 billion in notes. This increased the long-term debt to unprecedented heights of more than $7.0 billion and encumbered the corporation with a very large interest payment of over $900 million in 1982 and about $850 million in 1983. In order to reduce the debt, USS has been divesting itself of properties that were not consonant with overall objectives. These have been previously referred to and also include excess coal reserves, which were disposed of for several-hundred-million dollars. With this program, it is expected that the debt will be reduced much more rapidly than under normal circumstances.

The acquisition aroused considerable controversy, particularly with the United Steelworkers of America Union. Conscious of the number of jobs that have been lost in the steel industry during the past three years, the union maintained that USS, as well as other companies in the industry, should spend capital funds on steel plants and facilities to modernize them in order to improve their position as competitors in the world steel market. It was argued that the funds used to acquire Marathon could have been spent to build a new steel plant at Conneaut, Ohio, on Lake Erie. However, it must be pointed out that the $3.0 billion borrowed to consummate the Marathon transaction was more readily obtained since it was to be invested in a profitable oil company. On the other hand, it would have been very difficult, if not impossible, for USS to borrow $3.0 billion for the construction of a steel mill since the magnitude of the investment and the relatively poor prices obtainable for steel products would have made it a losing proposition. The return on a multibillion-dollar plant, with the steel price structure unstable and, in many instances, weak, would have made it virtually impossible to justify and obtain such a loan.

Marathon Oil has proven to be a significant asset in terms of profitability, which to a great extent reduces the corporation's steel losses. During 1982—which was by no means the best year for the oil industry—Marathon's sales amounted to $9.6 billion, and it generated an operating income of $1.238 billion. As a result of Marathon's profitable operation, the total loss for USS was $361 million, a figure far less than that which would have been registered had Marathon not been acquired, since the steel segment lost $852 million.

In 1983, Marathon's sales and operating income were down somewhat; sales were $9.3 billion and operating income was $1.1 billion. It still made a positive contribution. The acquisition of the oil company has done much to protect the steel company in times of slack steel demand, proving to be a cushion against the downward swing of the steel cycle. Further, when the debt and interest are substantially reduced in the next two to three years, it should make USS an attractive company in terms of profit.

Attempted Acquisition of National Steel

On February 1, 1984, USS made the dramatic announcement that it had entered into a preliminary agreement with National Intergroup to buy its steel facilities, formerly operated under the name of National Steel Corporation. This was a most unusual development, particularly since it involved the combination of the largest steel company and the seventh largest steel company in the country.

National Steel Corporation changed its name to National Intergroup in late 1983 after reducing its steelmaking capacity from 12.0 million tons in 1973 to less than 6.0 million tons in 1983. For some time, National's management was interested in divesting itself of its steel facilities, since it felt that the return on steel operations was too low to justify the additional investment needed to remain competitive. In 1981, National Steel reduced the size of its Great Lakes plant by 50 percent and in 1984 sold its Weirton plant to the employees.

It is interesting to note that when National Intergroup was formed in October 1983, a brochure was issued stating that the new company would maintain its core business, steel, which was the largest segment of its entire operation. However, when the opportunity to dispose of its steel operations presented itself, National Intergroup was quite willing to sell. The price ($575 million in cash plus the assumption by USS of a significant amount of debt) made a total of over $900 million. The consummation of this arrangement would have permitted National Intergroup to divest itself of all steelmaking operations, retaining only a chain of steel service centers that distribute steel.

From USS's point of view the acquisition of 5.5 million tons of steelmaking capacity for about $900 million, or less than $200 per ton, was highly advantageous. To construct these facilities in 1984 would have required an investment of more than $1,500 per ton. The facilities consist of two integrated steel plants and a modern finishing mill. One of the integrated plants is located near Detroit at Ecorse, Michigan, where National Steel produces sheets for the automotive industry as well as for other consumers. The other integrated plant is at Granite City (near St. Louis) and is a flat-rolled products producer with a diversified market. The Midwest Division of National Steel (near Chicago) is a modern finishing facility with cold-rolling as well as tinplate and galvanizing equipment.

Both integrated plants have some excellent facilities; each operates a modern hot-strip mill and continous-casting equipment. Steel is made at both locations by the basic-oxygen method; however, the blast furnaces at both plants require a considerable investment to put them in first-class condition.

There is no doubt that the acquisition would have placed USS in a much better position in terms of serving the U.S. durable-goods market, such as the automotive and appliance industries, since it would have acquired over 5.0 million tons of flat-rolled finishing capacity.

USS wanted more flat-rolled capacity, particularly since its facilities (which provide heavier products such as plates and structurals used for capital investment) have not been operating near their capacity due to sluggish market conditions. The recovery of 1983 and 1984 was felt in the consumer-durable-goods industries, such as the automotive industry, while investment in capital goods lagged. Thus, plates and structurals did not sell well, while the demand for sheets for consumer durables was extremely brisk.

The acquisition of National Intergroup's steel facilities would have allowed USS to participate to a greater extent in the consumer-durables segment of the economy.

National Steel (at its Great Lakes Division near Detroit) is short of coke, as is its Granite City plant, and this could have been supplied by USS through its Clairton Works (near Pittsburgh). The total coke shortage to be supplied its between 600,000 and 700,000 tons annually, depending on the rate of operations.

Further, National Steel has reduced its steelmaking capacity without a corresponding reduction in finishing facilities, so that it is short of steel to operate them fully. In a good year, it can roll 25 percent more steel than it can produce. Much, if not all, of this could have been supplied by USS; National would have had to purchase slabs outside of its own facilities.

Unfortunately for the two participants, the Justice Department announced that it would oppose the merger; however, the Department suggested a compromise: Two plants were to be sold before the merger could be consummated; the first plants suggested by Justice were the USS Geneva Works at Geneva, Utah, and the USS Pittsburg, California, plant. However, after making this offer, the Antitrust Division had second thoughts and required sale of the Fairless Works of USS and the Granite City plant of National. This meant that 6.0 million tons of capacity would have to be disposed of in order to gain 5.5 millions tons, and the only conceivable advantage to USS would have been a location in the Detroit market.

Both USS and National felt that the terms were too harsh, and they terminated their discussions. This was unfortunate for both companies and for the restructuring program of the U.S. steel industry. The two companies were a definite complement to each other. National was short of coke, which USS could supply. It was also short of melting capacity, since it had shut down one of the melt shops at the Great Lakes plant. This would have been reactivated by USS and would have provided melting capacity to operate the finishing facilities close to their potential and thus increase productivity and reduce costs.

Now that the merger has been denied, National will have to be in the market for 600,000 tons of coke and, at a high rate of operations, for several hundred thousand tons of semifinished steel. Much of this will be brought in from abroad, a situation which could have been avoided if the merger had been approved, since USS would have been able to provide National with most of the needed semifinished steel.

Notes

1. *1982 10-K Report,* United States Steel Corporation, p. 65.
2. *Press Release,* December 27, 1983, United States Steel Corporation, Pittsburgh, Pennsylvania.
3. *1982 10-K Report,* United States Steel Corporation, pp. 4–5.
4. *1982 Annual Report,* United States Steel Corporation, p. 4.
5. *1982 Annual Report,* United States Steel Corporation, p. 4.
6. *1982 Annual Report,* United States Steel Corporation, p. 4.

Bethlehem Steel Corporation

For more than sixty years Bethlehem Steel Corporation has been the second largest integrated producer in the United States. However, the merger between J&L and Republic coupled with the recent reduction in Bethlehem's steelmaking capacity will put Bethlehem in third place.

In the early 1970s, Bethlehem Steel had an annual raw-steel capacity of 25 million to 26 million tons but the loss of much of the West Coast market to Japanese imports brought with it reductions in capacity beginning with the Sparrows Point plant. Subsequent cuts in the capacity at Johnstown and Lackawanna in 1977 resulted in an even smaller steelmaking potential. This latter development was summed up in the 1977 Annual Report:

> During the year we announced the closing or curtailment of a number of operations and the attendant work force reduction. The biggest cutbacks began in August at which time we announced that the annual steelmaking capacity of the Lackawanna plant would be permanently reduced from 4.8 to 2.8 million tons and that the Johnstown plant annual steelmaking capacity would be permanently reduced by 0.6 million tons to 1.2 million tons.[1]

This reduction in capacity came shortly after the installation of a third basic-oxygen converter at the Burns Harbor plant. As a result of these changes, Bethlehem's raw-steel capability in 1977 was 22.2 million tons. In that year, Bethlehem also installed a second plate mill at Burns Harbor. This unit was 110 inches wide and added considerably to the plant's plate capacity.

In 1983, Bethlehem further reduced its capacity by closing the iron and steelmaking facilities as well as most of the finishing facilities at its Lackawanna plant. It also closed the Los Angeles plant, which was a relatively small electric-furnace operation. As a result, Bethlehem's raw-steel producing capacity as of 1984 stands at 17.5 to 18.0 million tons.

Bethlehem produces a variety of steel products, including hot- and cold-rolled as well as galvanized sheets and tinplate. It also has considerable capacity in the heavy products such as structural steel sections, plates, and rails.

Currently, it operates three integrated steel plants:

1. The Burns Harbor plant, near Chicago, which is the newest integrated, greenfield-site facility built in the United States with an annual capacity of 5.3 million tons.

2. The Sparrows Point plant, near Baltimore, which at one time was considered the largest or second largest steel mill in the world. It now has a capacity of approximately 7.0 million tons.
3. The original plant at Bethlehem, Pennsylvania, which is predominantly a producer of structural steel, with a raw-steel capacity of approximately 2.5 million tons.

The other major steel-producing plants, located at Johnstown, Pennsylvania and Steelton (near Harrisburg) Pennsylvania, operate electric furnaces and have a combined annual capacity of approximately 3.2 million tons. The company also operates a small electric-furnace plant in Seattle, Washington, which is currently for sale.

Bethlehem is not a diversified company to the extent that USS, Armco, and National are. Its interests and assets are predominantly in steel as stated in a recent petition submitted to the International Trade Commission. Bethlehem describes itself as: "[t]he nation's largest integrated, non-diversified producer of carbon and alloy (other than tool or stainless) steel mill products . . ."

In past years, the company operated some large shipyards for the construction of new ships. However, the principal ones have been sold and those that remain are dedicated principally to repair work. The construction division which had been prominent in the erection and construction of high-rise buildings and bridges was also eliminated.

Currently, Bethlehem, with its 17.5 to 18.0 million tons of annual steelmaking capacity, has set a goal of increased efficiency in production and selective marketing that will concentrate on those products that can be made and sold profitably. Others that have yielded an unsatisfactory return are being dropped. One product in which Bethlehem has a significant position is steel bars, and because these are a special quality item, the company does not feel minimill competition to any great degree. Nevertheless, it is a fact that the minimills are moving into the specialty bar field and may challenge Bethlehem to a greater extent in the years ahead. There is competition with minimills in concrete reinforcing bars which are produced in large tonnages in the Steelton plant.

The recent restructuring efforts that eliminated a number of operations are referred to in the 1982 Annual Report, which states:

> As part of the strategic plan, we began the process by closing down and disposing of a number of facilities and operations. Early in the year, we announced that we would be selling our ship repair yards at Baltimore, Boston, Hoboken, and San Francisco; all have been sold or are in the process of being sold. In September, we announced the closing of Bethlehem's Los Angeles plant and the intention to sell it and the Seattle plant. Presently, we expect to operate the Seattle facility until its sale.

Shortly before year-end, we announced a plan to restructure operations at the Lackawanna and Johnstown plants.[2]

Within the past decade, Bethlehem has also sold its plant with a galvanizing line for sheets at Pinole Point, near San Francisco. Further, in the early 1980s, the Johnstown plant was converted from an integrated coke-oven/blast-furnace/open-hearth operation to a modern electric-furnace plant. The Steelton plant—also an integrated unit with coke ovens, blast furnaces, and open hearths—was converted to an electric-furnace operation in the 1960s.

Capital Investment

Capital investment in 1984 and the years immediately beyond will be confined to essential improvements. Two continuous casters—one at Burns Harbor and the other at Sparrows Point—will be added to the two now in existence, one of which is at Burns Harbor and the other just completed at Steelton. The two casters that will be built by the Austrian company, Voest-Alpine, have been financed in a most unusual manner. A Bethlehem official stated, "This financing is unique not just because of its size, but also because of its terms."[3] The entire project, encompassing the two casters, is valued at $540 million, $40 million of which is interest that will be capitalized. Of this total, $350 million is being financed through American banks and the remaining $190 million through Austrain banks, $90 million of which is guaranteed by the Austrian government with the proviso that this will be paid to Austrian vendors. The owner-trustee of the two casters will be the Connecticut Bank and Trust Company, which will funnel the monies to Bethlehem which will then act as a general contractor for this construction. Bethlehem will lease the casters and make rental payments based on tonnage produced. The term of the lease is fifteen years, after which there is an option to acquire the equipment at approximately 20 percent of its original cost.

There are three groups involved in the financing:

1. Three American banks, having an equity participation (ownership banks), will provide $150 million.
2. Seven American lender banks, without equity participation, will provide $200 million.
3. Six Austrian lender banks, without equity participation, will provide $190 million of which $90 million is guaranteed by the Austrian government.

The financing arrangement is decidedly favorable since no money must be spent during the construction period.

With four continuous-casting units, Bethlehem will be casting 60 percent of its raw-steel output. The company is enthusiastic about its future, as expressed in a recent publication:

> Benefits of the huge capital outlay by Bethlehem for four continuous casters will accrue both to our customers and to ourselves. Customers will be getting products of exceptional metallurgy consistently with improved internal and external quality. Bethlehem will benefit from low energy costs, higher yield and increased productivity. Casters are win-win facilities despite their costs.
>
> Four continuous casters will represent an investment by Bethlehem of approximately $700 million over a 10-year period. This investment, even in a time when investment dollars were very tough to generate, was necessary if Bethlehem was to obtain its goal of becoming the most technically sophisticated, most competitive, and the best supplier of steel and services in the country.[4]

A major capital investment of approximately $50 million will be made to modernize the 48-inch structural mill at the Bethlehem plant. Another major capital investment will be required to modernize and upgrade the 68-inch hot-strip mill and a plate mill at Sparrows Point.

In terms of raw materials, Bethlehem draws upon ore mines and pellet plants in the United States and Canada. It has substantial interests in Minnesota where it participates in the Hibbing Taconite and Erie Mining Companies both of which have pellet plants. In Quebec it has approximately a 20 percent interest in the Iron Ore Company of Canada. Coal comes from its mines in Pennsylvania, West Virginia, and Kentucky.

Unlike the other major integrated steel producers, Bethlehem for the immediate future has decided to concentrate on steel rather than look for substantial diversification. The tone of its pronouncements indicated a confidence in the future of steel, provided such problems as imports can be solved. Bethlehem is determined to be an efficient producer capable of meeting competition from U.S. rivals as well as foreign producers.

Bethlehem Steel Works

At the Bethlehem steel plant, which was the original site of the company, the principal products are structural shapes and bars. The plant is fully integrated with six coke-oven batteries, four blast furnaces, a basic-oxygen steelmaking shop, and rolling mills. Several small electric furnaces for the production of alloy steel are also in operation.

The basic-oxygen shop can produce about 2.5 million tons of steel which are rolled principally into structural shapes.

The coke-oven segment of the plant is in good condition and will not require a major investment until at least 1990. At the present time, it produces a high stability coke and has an excess above the plant's requirements that will be sent to Burns Harbor.

The four blast furnaces will soon be reduced to three as the smallest unit will be abandoned when its present lining life gives out. In the future, the plant is conceived as a two-blast-furnace operation with a third furnace as backup when needed. It is in balance in terms of melting and finishing capacity and has no need to bring in semifinished steel from the outside. A modernization program is scheduled for the structural mill to maintain it as an efficient operating unit.

Burns Harbor

Burns Harbor is a fully modern integrated plant having been constructed in stages during the 1960s. It has two coke-oven batteries and two large blast furnaces, which are efficient and in every respect world-class. Each is capable of producing 6000 tons of iron per day. The basic-oxygen shop has three 300-ton vessels with a capacity to produce 5.3 million tons of steel.

Currently, the plant operates a continuous-casting machine for slabs and a second unit is under construction to be finished in 1986. At that time, 85 percent of the plant's steel will be continuously cast. When the second continuous caster with a capacity of 2.2 million tons is installed it will increase the yield at the plant from raw steel to finished product by 12 to 14 percent and will improve the plant's ability to provide aluminum-killed steel to the automotive and other industries. The increase in yield is needed since the plant can roll more steel on its 160-inch and 110-inch plate mills and the 80-inch hot-strip mill than the steelmaking shop is capable of producing. Heretofore, slabs were shipped from Lackawanna when needed. However, the abandonment of the Lackawanna plant has deprived Burns Harbor of this source of slab supply.

One of the principal finishing facilities at the plant is a continuous heat treating or annealing line for sheets. It has a capacity to heat treat 560,000 tons a year and represents an investment of $60 million. This unit replaces the traditional batch annealing furnace and performs the task much more rapidly and more effectively. It is one of two units in existence in the United States; the other is located at Inland Steel. With this facility (installed in 1983), Bethlehem will have high-quality sheets for its customers.

The Burns Harbor plant is short on coke, which can be shipped in from Bethlehem, Pennsylvania, where there is an excess of production, and from Lackawanna, where the coke plant is still operating.

In addition to the new continuous caster, no other major expenditures are planned.

Lackawanna

After sustaining losses from facilities that were not competitive, in late 1983, Bethlehem closed the blast furnaces, basic-oxygen converters, primary mills, hot-strip mill, cold-reduction mills, and structural mills, with a loss of approximately 3.0 million tons in steelmaking capacity. The closure, which terminated several thousand jobs, was not complete since the coke ovens that Bethlehem had wanted to sell will remain in operation throughout 1984 to provide coke for the Weirton Steel Company plant at Weirton, West Virginia. Further, the bar mill, a recent installation, as well as the galvanizing line for sheets, have been maintained in operation. Billets from Johnstown will feed the bar mill while the galvanizing line will be fed with cold-rolled steel coils from Sparrows Point. Just how long the coke ovens will remain in operation is a matter of conjecture; they will be operated while there is a demand for coke and no buyer comes forth to purchase them.

Sparrows Point

Sparrows Point is Bethlehem's largest plant with a steelmaking capacity of approximately 7.0 million tons. Currently, there are seven operable coke oven batteries, among which is a new unit put in operation in 1982 that replaced three older batteries. There was a plan to replace six of the current seven batteries with a large battery; however, this has been postponed indefinitely since it would represent an investment of over $250 million, which at the present time does not have a high priority.

Sparrows Point has an adequate coke supply for its four blast furnaces. Three of these—designated H, J, and K—are approximately the same size, with each capable of producing from 3,000 to 3,200 tons of iron per day. Furnace L is a new, large, world-class furnace brought on in 1978 with a 10,000-ton-a-day capacity. The operating plan is to produce iron at L furnace and, when business warrants it, at one of the other three. Presently, the L furnace is operating with the other three banked. However, when it is necessary to take L furnace down for a complete relining, the other three furnaces can be put back in operation. In the fall of 1984, L furnace is scheduled to undergo minor repairs and will be out of operation for perhaps four weeks. At that time, one of the other three furnaces may be brought on to fill the gap.

Steelmaking equipment at Sparrows Point consists of two 220-ton basic-oxygen converters installed in 1966. These will be replaced in October 1984 by two 280-ton vessels and the capacity of the shop will be increased to 3.6 million tons. To allow the new units to function at their full potential, adjustments will be made to the teeming aisle and the overhead cranes.

In addition to the BOF shop, there are seven 420-ton open-hearth furnaces that were put in operation in 1958; this represents the last open-hearth shop built in the United States. When the demand for steel is high, six of these can be operated at one time, giving the shop a potential of 3.7 to 4.0 million tons. However, it is more probable that its production will be in the area of 3.0 to 3.5 million tons.

A new continuous casting machine is under construction at Sparrows Point with a capacity of approximately 3.0 million tons. It will be completed in 1986 and will increase the yield from raw steel to finished product for the tonnage involved by 12 percent to 15 percent.

The plant operates two plate mills and two hot-strip mills. The strip mills are 68 inches and 56 inches wide and, as indicated, the 68-inch mill will be modernized in the future, after which the 56-inch mill probably will be shut down. One of the plate mills will also be modernized, while no investments are needed for either the cold-rolling or finishing facilities.

Johnstown and Steelton Plants

The Johnstown and Steelton plants will continue to function as electric-furnace operations. At Steelton, a new continuous caster has recently been installed that will process the entire output of the plant. Finished products are rails and concrete reinforcing bars.

At Johnstown, the two large electric furnaces with a capacity in excess of 1.5 million tons per year will provide steel for the bar mills at that plant as well as billets for the bar mill at Lackawanna.

Future Prospects

Bethlehem, with its restructuring almost complete, will produce steel in its five major plants as well as one small plant in Seattle until it can be sold. Steel and coke will be shipped between plants as demand requires which puts the company in balance although some of its plants are not.

By 1986, with new continuous-casting capability in operation, and obsolete plants eliminated, it should be competitive in terms of quality and costs. It will be a lean, streamlined company that emphasizes steel production. The objective for the future was stated in the 1982 Annual Report by Donald Trautlein, Chairman of the Board. In speaking of the strategic plan for the future, he said:

> The objective of this plan is to make Bethlehem the best steel producer in this country in terms of productivity, quality, service, and profitability. To achieve that goal requires change and Bethlehem emerged from 1982 as a changed company.

> We've completed the reorganization of most of Bethlehem's operating activities . . . the most significant is the Steel Group comprising over 75% of Bethlehem's business.

The 75 percent concentration in steel indicates Bethlehem's future course. It is interesting to note that included in the 17.5 to 18.0 million tons of steelmaking capacity is the open-hearth shop at Sparrows Point. This type of facility is considered obsolete by many and requires more time for the production of steel than the basic-oxygen converter. However, with the application of oxygen to the open hearth, productivity has been improved. Further, the open hearth offers a degree of flexibility in utilizing scrap and hot metal far beyond that of the basic-oxygen converter. It can be a low-cost facility if the price of scrap falls; however, if scrap prices are high, operating costs will increase accordingly. Further, replacement with an oxygen converter would require a considerable capital investment, which the company does not feel is justified at this time. This open-hearth shop—the last built in the United States—will remain a production unit through most of the 1980s.

Bethlehem, as indicated, is not as diversified as many other steel companies, and diversification is not a top priority. However, should an opportunity arise, the company would have to give it serious consideration.

Notes

1. *1977 Annual Report*, Bethlehem Steel Corporation, p. 3.
2. *1982 Annual Report*, Bethlehem Steel Corporation, p. 1.
3. *Iron and Steelmaker*, February 1984, p. 44.
4. *Bethlehem Review 1983/4*, p. 5.

Inland Steel Company

Inland Steel Company, a fully integrated producer, is overwhelmingly committed to steel and steel-related activities. In addition to its integrated steel mill, its other divisions are very much involved with steel. They consist of:

1. The largest steel service center combine in the United States, Joseph T. Ryerson and Son, Inc., with twenty-five outlets.
2. Inland Steel Container Company, a manufacturer of steel pails and drums.
3. Inryco, Inc., a leading supplier of steel and other building products.

Another division that is not particularly steel-related is Inland Steel Urban Development Corporation, engaged in housing manufacture and real-estate development. Inland also has a joint venture with A.O. Smith, which manufactures fiberglass reinforcing plastic pipe systems.

The integrated steel operation, as well as the three steel-related divisions, account for more than 95 percent of Inland's sales.

Inland Steel Company is unique among the *major* integrated steel producers insofar as it operates only one plant. As a consequence, all of its efforts and investments are concentrated at the one facility at East Chicago, Indiana. It is the largest in North America with an ironmaking capacity in excess of 8.0 million tons and a steelmaking capacity of 9.3 million tons. The plant's facilities consist of a number of coke-oven batteries, nine blast furnaces, and a variety of steelmaking facilities, including basic oxygen converters, electric furnaces, and an open-hearth shop. There are two continuous casters currently in operation and a third is under construction. Finishing facilities are predominantly devoted to light flat-rolled products. There are three hot-strip mills and one plate mill, as well as four bar mills and one structural mill. The hot-strip mills supply cold-reduction mills and coating facilities for the production of hot-rolled and cold-rolled sheets, as well as galvanized sheets.

At full operation, Inland has more steelmaking capacity than finishing capacity and until adjustments can be made—either through the construction of additional facilities, which seems highly unlikely at present, or an arrangement with another steel company—Inland will not be able to produce up to its steelmaking potential.

Inland is not a diversified company and remains heavily committed to the production of steel. An examination of its annual reports over the last decade bears this out, since they are almost completely devoted to steelmaking and marketing.

In 1974, Inland announced a huge capital investment program which ultimately required more than $1 billion. It was intended, among other things, to increase steelmaking capacity from approximately 8.0 million tons to 10.5 million, allowing the company to ship 8.0 million tons of finished product. The units to be installed included:

1. A new blast furnace with an ultimate potential capacity of 10,000 tons a day.
2. A coke-oven battery with pipeline charging which represented the latest technology.
3. A 160-inch plate mill with a 600,000-ton annual capacity.
4. A continuous caster for slabs.
5. A renovation of the 14-inch bar mill.

In addition, raw-material supplies of iron ore and coal were augmented.

Except for the plate mill, the program was completed and the facilities put in operation in 1980. The plate mill, a casualty of increased costs of installation as well as the deterioration of that segment of the steel market, was dropped. When the plan was abandoned, the projected cost of the plate mill and its finishing facilities had risen to $400 million. Further, surveys of the plate market were not particularly optimistic. The plate mill cancellation left the plant at Indiana Harbor out of balance, since it can produce more steel than its current finishing facilities can absorb. A 1982 attempt to establish a joint venture with Armco, which would have given Inland access to more finishing facilities, was unsuccessful.

Inland has a plentiful supply of raw materials, with participation in a number of taconite mines and pellet plants in the United States and Canada, as well as ownership and investment in coal mines in Illinois and Pennsylvania.

In terms of corporate changes, Inland has not experienced the restructuring that other companies have undergone. It has concentrated its efforts on strengthening customer relations by improving steel quality, introducing new products, and expanding sales coverage. As of 1982, its market was confined principally to the Midwest as indicated in the 10-K report for that year: "Approximately 95% of the company's shipments of steel mill products in 1982 were to customers in 20 mid American states and approximately 72% were to customers in a five-state area comprised of Illinois, Indiana, Ohio, Michigan, and Wisconsin."[1] In this geographic area, Inland competes not only with the nation's largest producers of steel but with many smaller mills, as well as foreign producers who ship their products into the Midwest.

In 1982, the company adopted a plan to expand its sales coverage which will involve greater efforts to market steel in the South and Southwest where new sales offices were opened.

Inland is a major supplier of the automotive industry. In 1981, it shipped a total of 5.8 million tons, of which 24 percent went to transportation, which is dominated by the automotive industry. In 1982, Inland's shipments fell to 4.1 million tons; however, 30 percent went to the transportation industry.

Recent and current investments in the plant at Indiana Harbor include a continuous annealing line for sheets, which was put in operation in 1983. This was a very important installation and is one of two such lines in the entire U.S. steel industry. The other is at the Burns Harbor plant of Bethlehem Steel. The unit, which has a 400,000-ton-per-year capacity, greatly reduces the time required for annealing and produces high-strength sheet steels with special formability which are used by the automobile and appliance industries.

Another major installation of great significance is the third continuous-casting unit, which will have a capacity to process 2.2 million tons and will be in operation in 1986. At that time, Inland will be able to cast 4.5 million tons or half of the plant's potential output. This calculation is based on a 100 percent operating rate. At lower rates of production, continuous casting will account for a higher percentage of output. An unusual financing arrangement has been made for the continuous caster, which does not require Inland to begin paying for the unit until the start of production.

Like Bethlehem, Inland continues to operate an open-hearth furnace shop. This consists of seven 335-ton furnaces, and in 1981 and 1982, it accounted for 21 percent and 13 percent respectively of Inland Steel's production. The furnaces are equipped with oxygen lances and produce a heat of steel in approximately five hours. They afford the plant a degree of flexibility insofar as over 50 percent of the charge is in the form of scrap. When this material is available at a low price, the operating costs are favorable. It seems certain that the open-hearth shop will continue to function for the remainder of the decade.

Unlike virtually all of the integrated producers, Inland has not shut down permanently any of its major manufacturing units at Indiana Harbor. Operations at a number of them have been suspended because of market conditions; however, they are on standby and will be brought back when steel demand improves. Likewise, unlike most of the major integrated producers, Inland has not reduced its steelmaking capacity and has no plans to do so. The current excess of iron and steel production over and above rolling capabilities could lead to an arrangement that might be made for Inland to provide iron to the Indiana Harbor Works of J&L, which is short on blast-furnace capacity. The plants are located adjacent to each other and transportation costs would be negligible.

One problem that Inland faces at its Indiana Harbor plant is a shortage of coke. When the new large blast furnace was built, there were plans to add another battery of coke ovens; however, the estimated capital cost of $250 million led to the indefinite postponement of this project. Further, one of Inland's recently installed coke-oven batteries, which encountered operating problems, has been shut down permanently, thus aggravating the coke shortage. As a consequence, it is necessary to purchase coke from outside sources.

Inland's complete commitment to steel means that the 9.3-million-ton capacity at Indiana Harbor will most probably be kept intact. There may be some adjustments to blast furnace facilities, but this will merely be a shift in production rather than a cut. The company will continue through improving its steel quality to strive to meet the more demanding needs of the customers. The continuous-casting operations, as well as the introduction of ladle metallurgy, will be of significance in this program.

Another manifestation of this determination is the recent announcement made jointly with Bethlehem Steel to research and develop ways of improving the production of galvanized steel. This is important to both companies since they are large suppliers to the automotive industry where a notable shift is being made to galvanized steel to improve corrosion resistance in the automobile body.

The future of the company is summed up in a comment in the 1982 Annual Report:

> We believe (and qualified outsiders agree) that Inland is among the best situated steel companies for upside potential once normal steel demand returns. The expansion and modernization programs of the 1970s and our current efforts at margin and customer performance improvements, all are directed at achieving this end to the benefit of Inland's owners as well as our employees.[2]

Notes

1. *1982 10-K Report*, Inland Steel Company, p. 2.
2. *1982 Annual Report*, Inland Steel Company, p. 4.

Armco Steel Corporation

Armco is a diversified steel producer with somewhat less than one-half of its sales in steel. This development toward diversification began in 1958 with the acquisition of National Supply Company, which at that time was one of the world's largest manufacturers and distributors of oil-field machinery and equipment. In a sense, this was not a complete break with the steel industry, since National Supply produced oil-country tubular goods as well as oil field machinery and equipment. In all of its operations, National Supply was a large consumer of steel and provided Armco with a substantial market for its steel output.

A further step toward diversification was taken in 1966 when Armco formed a strategic planning task force whose purpose was to study new markets, products, and processes both in the United States and abroad. The annual report for that year stated:

> In a world in which accelerated technology can influence a company's operations so profoundly, a company which doesn't seek out new ideas and new visions cannot grow for the maxium benefit of its shareholders.[1]

The committee's activity resulted in the acquisition in 1969 of the Oregon Metallurgical Corporation, a producer of titanium. In 1968 Armco set up an equipment-leasing operation—Armco/Boothe Corporation—which was designed to provide financial services to industry. In 1969, Hitco, a producer of nonmetallic composites, was acquired. This gave Armco a strong position in the field of nonmetallic composites, many of which had applications in aerospace technology.

As a consequence of the program, which was started in 1958 and has continued through the 1980s, Armco is truly a diversified company. At present the divisions are:

1. Steel, divided into carbon and specialty.
2. Oil-field equipment.
3. Aerospace and strategic materials.
4. Financial services, which include not only leasing but also insurance.
5. Fabricated products and services.

Also, Armco has operations in twenty-six foreign countries.

Despite the acquisitions, during the late 1960s Armco was still committed heavily to the production and sale of steel.

In 1964, it launched Project 600 which involved an investment of $600 million over a six-year period to improve and expand steelmaking facilities. By 1967, it was evident that the goals set for this project could not be reached without an additional expenditure of $200 million. Some of the principal installations included in the project were a basic-oxygen steelmaking shop, an 86-inch hot-strip mill (the widest in the United States), as well as a wide cold-reduction mill. These facilities were constructed at the Middletown plant, while at the Butler Works in Pennsylvania, which was devoted to specialty steel, three electric furnaces were installed to replace open-hearth furnaces.

Steel operations, which in the 1970s constituted Armco's principal activity, are divided into carbon steel and specialty steel. Carbon steel plants are located at Ashland, Kentucky; Middletown, Ohio; Kansas City, Missouri; and a small electric furnace plant in Mexico City, Mexico. During the past six to seven years, Armco has steadily decreased the portion of its investment in carbon steel but has maintained a significant group of assets in the specialty steel segment of its business. The principal specialty steel plants are located at Butler, Pennsylvania, and Baltimore, Maryland.

In the 1981 Annual Report a strategic plan was outlined, indicating the path the corporation would follow in the process of diversification. Table 6–1 gives the distribution of assets in 1976, 1981 and the goal for 1985:

Table 6–1
Distribution of Assets. Years: 1976, 1981, and the Goal for 1985

	Percentage		
	1976	*1981*	*1985 Goal*
Oil-field equipment and production	10	23	35
Aerospace and strategic material	1	9	9
Specialty steels	10	7	7
Financial services	3	8	8
Fabricated products and services	12	10	12
Carbon steel	64	43	29
	100	100	100
Overall Net Assets (*billions of dollars*)	$2.4	3.6	5.8

This plan set forth in 1981 called for significant growth in oil-field equipment. At that time, the oil business was booming with a record number of drilling rigs in operation. To maintain its market position, Armco planned an expansion of its oil-country goods facilities to achieve a production of 750,000 tons of this product. A new mill with a capacity to produce 450,000 tons of oil-country tubular goods was planned at an estimated cost (with auxiliaries) of $670 million. This was considered an attractive investment in light of the oil industry activity at that time. However, in 1982, the oil drilling boom collapsed. Drilling rigs in operation dropped from 4530 in December of 1981 to below 2000 by mid-1982. Further, orders for oil country tubular goods fell sharply in 1982 and continued their decline in 1983. Faced with these realities, Armco first postponed its plan to build the tubular mill and later cancelled it after an attempt to organize a joint venture failed. Thus, the 35 percent in assets of oil-field equipment will not be realized by 1985.

The diversification that Armco has developed over the years has been helpful in terms of maintaining earnings. The management stated in 1978 that the "drive to transform Armco into more than a steel company" was very successful in terms of maintaining earnings.

In 1983 Armco reduced the size of its steelmaking capacity with the sale of two minimills, one at Sand Springs, Oklahoma, and the other at Marion, Ohio, and at the end of the year the closure of the Houston plant with 1.0 million tons of capacity, a development brought on by drastic losses incurred at that plant. Thus, in terms of carbon steel, the major operations in 1984 consist of two fully integrated plants: one at Ashland, Kentucky, and the other at Middletown, Ohio, as well as the Kansas City plant based on electric furnaces and scrap.

In order to streamline and make the carbon steel operation more efficient, the two integrated plants will be coordinated as far as possible, considering that there is a distance of about 130 miles between them. These developments are in contrast to the commitment to carbon steel made in the middle to late 1960s when Armco carried out its Project 600. In 1982, there were sales of coal properties (not needed) for $300 million. Also in 1982, with the restructuring of the corporation, some 15,000 employees were dropped.

Middletown Works

Middletown is the original location of Armco that started at the turn of the century. Today the plant consists of three coke-oven batteries, one blast furnace, a basic-oxygen shop, and an open hearth, as well as hot and cold strip mills. The blast furnace is not adequate to produce enough iron for the steelmaking facilities, so its output must be supplemented by two small blast furnaces operating at Hamilton, Ohio, some twelve miles away. These two furnaces until 1982 were fed by two coke oven batteries which were abandoned at that time. The furnaces have an aggregate output of about 2300 tons per day and will be rebuilt in the next two years.

The original intention of the company was to abandon these furnaces at the end of their useful life, which was calculated to be 1985, and replace them with a 5500-ton-per-day modern blast furnace at Middletown. However, the cost of such an installation was estimated in 1983 at $380 million, and Armco's Board of Directors decided against this investment. Thus the two furnaces at Hamilton will be relined and given new shells that will add a small amount to their hearth diameter. This operation will be undertaken in 1984 and 1985, and over a ten-year period (considering periodic relinings) the entire cost will be in the area of $45 to $50 million. The larger furnace at Middletown will be relined in 1984 with considerable improvement to its stoves to permit blast temperatures of 2000 degrees and will result in a daily production of 4000 to 4200 tons of iron.

The Middletown steelmaking facilities consist of two 225-ton basic oxygen vessels installed in 1969 and six open-hearth furnaces. Because of the limited amount of hot metal, the open hearths are operated on a 90 percent scrap-pig iron and 10 percent hot metal charge. With four of the six furnaces in operation, the capacity of the shop ranges from 1.1 to 1.2 million tons annually while the basic oxygen converters have a capacity of 2.2 million tons.

The Middletown plant operates a continuous caster for the production of slabs which is functioning well beyond its originally designed capacity and processes about one-third of the steel produced at the plant. The remainder is rolled on a slabbing mill which was installed along with the continous hot strip mill in the 1960s. By the end of 1985 the company plans to close the open-hearth shop, thus reducing steel output.

The finishing facilities at the Middletown Works include a modern hot strip mill and several cold reduction mills. The 86-inch-wide hot-strip mill is the widest modern mill in the United States. If the steel were available it could readily produce 4.5 million tons of hot-rolled coils. However, there is not enough steel produced at Middletown to bring the mill up to its full potential.

The Middletown plant currently has an excess of coking capacity and a shortage of steel in relation to its rolling potential.

Ashland Works

The plant at Ashland, Kentucky, is historically important since the steel industry's first continuous hot-strip mill was built there in the early 1920s. The plant currently is fully integrated since Armco purchased two batteries of coke ovens from the Allied Corporation in late 1982. These are located close by and feed the two blast furnaces at Ashland. One of the furnaces, designated Bellefonte, was relined in 1981 with considerable improvements including three new stoves. The furnace has a production capability of approximately 2800 tons of iron per day. The second furnace, designated Amanda, was taken out of operation in November 1983 and will undergo significant repairs as well as a relining. The stoves will be improved, allowing higher temperatures and an increase of furnace production to 4200 tons per day. Ashland's steelmaking capacity from its two 215-ton BOFs is approximately 2.2 million tons. Iron from the two blast furnaces is more than enough to care for the steel shop's requirement. With the Amanda furnace in operation, in 1984 there will be an excess of iron, some of which can be shipped in the form of cold pig iron to Middletown for use in the open hearth.

A continuous caster has just been constructed at the Ashland Works which will be versatile insofar as it will cast both blooms and slabs. Originally this was intended to be a bloom caster to provide material for the new 450,000-ton seamless pipe mill that Armco had projected and hoped to have in operation by 1984. As indicated, this project has been abandoned. Thus the bloom caster was converted into a combination bloom and slab caster. Blooms are provided for the existing seamless mill at Ambridge, Pennsylvania, and slabs for the strip mill at Ashland. Like Middletown, Ashland is a light-flat-products mill with finishing facilities consisting of hot- and cold-strip mills.

The integration of the two plants that is now under consideration will probably result in the transportation to Middletown of some slabs cast at Ashland, thus supplementing the inadequate steel supply at the Middletown plant. These will be converted into hot-rolled coils, some of which may be sent back to Ashland for cold reduction.

For the strip mill at Middletown to operate at an efficient 4.0-million-ton rate or more per year, several hundred thousand tons of slabs from Ashland will be needed, and this will probably result in a closure of Ashland's hot-strip mill.

The future coordination of the two plants will also involve a transfer of coke. The present coke ovens at Ashland can provide about 1.0 million tons annually; this is not enough to feed the two blast furnaces, if they are to make 2.5 million tons of iron to feed the BOF at Ashland and the open hearths at Middletown. The coke supply will be supplemented from Middletown, where Armco has a production of more than 1.5 million tons, which is in excess of its blast-furnace requirements at both Middletown and Hamilton. These transfers of coke and steel will tie the two plants together to increase efficiency and reduce costs.

Kansas City, Missouri

The Kansas City plant can be considered as two. It has two electric-furnace shops: one installed in the late 1970s with four 150-ton furnaces and a bloom caster, and another, which is much older, with two 125-ton furnaces.

The products of these plants are bars, wire rods, and wire. It is quite conceivable that in the near future the older furnace shop may be shut down and all of the steel produced by the new electric furnaces.

Specialty Steel

Armco Steel will continue to improve its specialty steel operations in the two plants located at Butler, Pennsylvania, and Baltimore, Maryland. The Butler plant is a large producer with three 165-ton electric furnaces and continuous casting for the entire output. It produces a variety of specialty steel products in sheet form and has a steelmaking capacity of approximately 1.0 million tons annually. The Baltimore plant is devoted to stainless steel products and operates two electric furnaces with 35 and 50 tons capacity, respectively.

The specialty steel segment of the business will continue to receive considerable attention in the years ahead.

Armco has reduced the size of its steelmaking capacity in the carbon steel area by writing off or selling capacity that was considered either noncompetitive or not in keeping with its future plans. By 1985 carbon steelmaking capacity will be approximately 6.0 million tons with more than 1.0 million tons of specialty steel. It is virtually certain that there will be no additions to these amounts in the foreseeable future.

With improvements to the existing facilities—particularly the iron and steelmaking equipment at Middletown and Hamilton—as well as the improved blast furnace operation at Ashland, the production of carbon steel should be carried out on an efficient basis. There may be some further reductions in capacity, such as the open-hearth shop at Middletown.

The steel plants are backed up by adequate supplies of raw materials consisting of 50 percent participation in Reserve Mining Company as well as interests in the Iron Ore Company of Canada and Eveleth Expansion Company. Coal is provided from the company's mines in Oklahoma and West Virginia.

The trend toward diversification will continue whenever opportunities present themselves. The latest was the acquisition in 1981 of the Ladish Company, a leading manufacturer of footings, fittings, and valves.

Most of Armco's current restructuring has been accomplished, and within the next year or two with the adjustments at Middletown and Ashland, the task will be completed.

Note

1. *1966 Annual Report,* Armco Steel Corporation, p. 3.

National Intergroup

The new name adopted by National Steel Corporation in October 1983 represents its drive to diversify. This started in earnest in 1967 when National began to study the possibility of entering the aluminum industry. It took the step in 1968 by acquiring 20 percent of Southwire Corporation in Georgia and, along with Southwire, acquired a 50 percent ownership in an aluminum smelter. Several aluminum fabricating companies were added in 1970.

In the early 1970s, although the company was somewhat diversified, it was still very much inclined to expand and develop its steelmaking operations.

In the midst of the steel boom in 1974, National Steel announced the expansion of its Midwest plant which was to be completed in 1980. The expansion was aimed at integrating the existing finishing facilities at Midwest which had been in operation since 1961. The 1974 Annual Report stated under heading of "Expansion—A New Steel Mill."

> Believing as we do that steel demand will move up in the years ahead at a higher rate than in the recent past, we must expand our steelmaking capacity to meet the resulting requirements.[1]

National subscribed to the then-current thinking that steel demand would grow at a rate higher than 3 percent per year, and by 1980 there would be a need for 30 million more tons of capacity than existed in 1973. In order to maintain its position in the market, the company decided to add 2.3 million tons of steelmaking capacity at the Midwest facility. The installation was to consist of a battery of coke ovens, a 5,000-ton-per-day blast furnace, a basic-oxygen steelmaking shop that could produce 2.3 million tons, a continuous caster, and a hot-strip mill. Since finishing facilities were already installed, as well as port facilities and other infrastructure, the financial requirements would not be as great as those needed for a greenfield site plant. Work on the engineering aspect of the plant proceeded through 1975 and part of 1976. However, it should be noted that at the outset National expressed caution in respect to the erection of the plant. In the 1974 Annual Report, in connection with the financing of the new plant, National made these remarks:

> Our projections indicate that these total financial requirements are well within our capabilities, but this is directly dependent on our being able to

sustain reasonable financial returns on our sales over the next five years. Should we incur or should governmental policies seek again to restrict earnings to an unreasonably low level, major segments of the program will be in danger and need to defer them could arise.[2]

The original estimate for construction of the plant was in the area of $1.1 billion.

The sharp drop in steel activity in 1975 and its failure to recover in 1976 and 1977 caused National to reconsider its position. In the 1976 Annual Report, just two years after the announcement, National's decision to postpone its construction of the Midwest plant was announced. The report stated:

We have felt it desirable to defer our Midwest expansion program because of continuing levels of low profitability resulting in a reduced cash flow. Generation of the substantial capital outlays required for this program is not yet in sight from internal sources. We have borrowed funds both to carry out present expansion as well as to replenish working capital. These have increased our long-term debt and we do not contemplate at this time additional borrowing to carry out this major expansion move. The overall Midwest expansion program will continue under active review for implementation when circumstances permit. In addition to the need for borrowing to finance the program it was evident that the costs of installing the facilities at Midwest would be considerably higher than the original estimate.[3]

As a result of these conditions, in the next two years, the project was quietly abandoned.

The years 1976 through 1979—although reasonably profitable in terms of steel production—were not judged to be adequate by National's management to pursue development in steel. In 1979, the company decided to diversify and in January of 1980 acquired United Financial Corporation of California for $241 million, thus establishing itself in the business of financial services. In 1981, an additional financial company was acquired. The Annual Report for 1979 stated:

The acquisition of United Financial Corporation of California by National Steel . . . marks the first entry by the Corporation into the financial services field in National's continuing diversification program.[4]

This annual report also stated:

The overall strategy of our Company has included, as a major thrust, diversification into other lines of business which either (a) complements or supports our business or (b) afford us the opportunity to grow in totally unrelated lines where prospects for profitability are promising. We propose to continue to emphasize that effort.[5]

The trend to diversification as well as the reduction of steelmaking capacity continued in 1981, when a decision was made that reduced the size of the Great Lakes plant (near Detroit). In that year, the Number 1 basic-oxygen steelmaking shop was closed down, and the blast furnace activity reduced from four to two furnaces. This represented a loss of 3.0 million tons of steelmaking capacity, leaving the plant with somewhat in excess of 3.0 million tons, produced mainly by the Number 2 BOF shop and, to a lesser extent, by the electric furnaces.

A further reduction in steelmaking capacity was decided upon in 1983 when National made a proposition to its employees at the Weirton plant in West Virginia that they buy the plant and operate it themselves. National's 10-K Report to the Securities and Exchange Commission for 1982 stated:

> In 1982, the registrant made a decision to substantially limit its future capital investment at its basic steel-producing plant at Weirton, West Virginia as part of the registrant's overall goal for directing its capital funds to areas of highest return. As a consequence the registrant offered to sell the principal assets of the Weirton Steel Division to an employee group, and the registrant has been engaged in negotiations with union representatives and Division management with respect to such sale.[6]

In March of 1983, an agreement was reached in principle for the sale of the facility to the employees. The final transaction was completed in January 1984 and the Weirton Steel Division of National Steel became an independent company to be known as Weirton Steel Corporation.

As a result of the divestiture of much of this capacity, National's ability to produce steel dropped from almost 12.0 million tons per year in 1973 to less than 6.0 million tons in 1984. Nevertheless, steel remained the core of National's newly reorganized Intergroup, whose divisions consist of:

1. Steel production.
2. Financial services.
3. Distribution through a chain of service centers.
4. Energy including National's mines.
5. Aluminum production.
6. Diversified businesses including pipe and tubing as well as some fabrication.

The brochure issued announcing the formation of the new group placed a great deal of emphasis on its steel operation. It stated in relation to steel, "As National Intergroup, we remain strongly committed to our core business—steel. Our steel operations continue to carry the name National Steel Corporation."[7] The brochure goes on to state that, although reduced in size, the emphasis will be placed on target markets and the customers they represent. It states, "National is a market-driven, customer-oriented steel supplier, intent on becoming *the* quality, low-cost producer of sheet steel products."[8]

With the sale of the Weirton plant, steelmaking is now limited to two integrated plants: one located at Granite City, Illinois, and the other at Ecorse, Michigan, near Detroit. The company operates finishing facilities including cold-reduction equipment as well as facilities for tinplating and galvanizing at its Midwest plant near Chicago. Steel production, although much smaller, is still the dominant operation in the company's activities. The steel plants have undergone considerable changes in the past two decades, and there will be some further changes in the years ahead.

Great Lakes Plant

The location of this plant near Detroit indicates its service to and dependence on the automotive industry for a market. To serve the automotive industry, the Great Lakes plant is almost exclusively a producer of various types of steel sheets. The steelmaking facilities at Great Lakes consist of two basic-oxygen shops and two 150-ton electric furnaces. In the rolling and finishing section of the plant, there is a continuous casting machine, an 80-inch hot-strip mill, and an 80-inch cold-reduction mill.

The Number 1 BOF shop, installed in 1962 with a capacity of 3.0 to 3.5 million tons, has been permanently shut down. Steel is made in the Number 2 BOF shop with an annual capacity of approximately 2.6 to 2.8 million tons. The two electric furnaces supplement this production with a total annual capacity of about 500,000 tons. There are four blast furnaces standing at the plant, only two of which are operating. The third is used as a standby unit, and the fourth has been abandoned. The two operating furnaces are due to be relined in the next two to three years.

The plant currently operates two batteries of coke ovens, one of which is scheduled, because of EPA regulations, to be shut down. However, it will operate until the other has been rebuilt from the pad up in 1986 or 1987. Thus the Great Lakes plant will have only one battery and will be considerably short of coke to meet its requirements.

The plant has a continuous casting operation that processes most of the steel poured from the basic-oxygen shop. Very serious consideration is being given to a second continuous caster which will allow Great Lakes to cast 100 percent of its steel output. In addition, an electrolytic galvanizing line is planned to meet future demands of the automobile industry.

Since closing the Number 1 BOF shop, Great Lakes has much more finishing than melting capacity, so that its hot-strip mill and cold-reduction mill can process up to 25 percent more steel than the plant can produce. Consequently, in times of strong steel demand, the plant will have to bring in semifinished steel.

Granite City

The Granite City plant was the sole producing unit for Granite City Steel Corporation, acquired by National in a merger in 1971. It is a fully integrated plant with a steelmaking capacity of 2.1 to 2.2 million tons annually. Currently there are three coke-oven batteries standing, two of which have recently been rebuilt and are in operation. The third is not operating and will be abandoned. Consequently, Granite City will be short of coke to feed its two blast furnaces, which provide iron for the basic-oxygen steelmaking shop. About 50 percent of the steel from this shop is fed through a continuous caster installed in 1983. The plant's finishing facilities consist of an 80-inch-wide, modern hot-strip mill, and a cold reduction mill. These facilities are capable of processing more steel than the plant can produce, so in times of high steel activity, slabs have been brought in to be processed.

Granite City—unlike Great Lakes which is closely tied to the automotive industry—supplies a diversified market. There are no plans to increase the capacity of this plant in the near future; however, considerable work must be done on one of its blast furnaces in the next year or two. As indicated, output can increase if semifinished steel is brought in from outside.

Granite City is short of coke as is the Great Lakes plant, and at an operating rate of 80 to 85 percent, the amount needed to make up the deficiency in both plants is close to 700,000 tons.

Midwest Plant

The Midwest plant of National was put in operation in 1961. It is a finishing facility fed by hot-rolled bands, which it turns into cold-rolled sheets, tinplate, and galvanized sheets. This plant depended on the other three National Steel plants for its steel. However, the reduced capacity at Great Lakes and the disaffiliation of Weirton make it difficult to supply the needs of Midwest from within the corporation. To remedy the situation, National has agreed with Weirton to purchase 180,000 tons of hot-rolled coils per year for the next five years. Partial payment for this will be an exchange of ore, which will come from National's participation in the Iron Ore Company of Canada.

In the last few years, National has closed its blast-furnace operation, as well as its coke ovens, in the Buffalo, New York, area, which were under the management of the Hanna Furnace Company. This operation was a pig iron producer, with the coke ovens operated as a joint venture with Republic Steel.

National has carried out a considerable shrinkage in its steelmaking capacity, coupled with a diversification into a number of nonsteel areas. In this it is not unlike some of the other major integrated steel producers. As a consequence of its development during the past few years, National has become a smaller yet, with the installation of continuous casting, a more efficient operation. Its dependence on the automotive industry at the Great Lakes plant is extremely heavy, and to a great extent, its fortunes in the steelmaking operations are linked to those of the automotive industry. For a number of reasons discussed elsewhere, the automobile industry will require less steel than it did in the 1970s, and the realistic appraisal of this position has led National Steel to reduce the size of its Great Lakes plant. National will, however, be in a position to provide significant tonnages of higher-quality steel to the automobile producers through the remainder of the 1980s.

National's reduced steelmaking facilities has also led to the sale of excess coal reserves. Some of these reserves were disposed of in 1982, with the Energy Group of National Intergroup mounting a concerted effort to market its coal.

Sale of Steel Assets

On February 1, 1984, National Intergroup and USS announced that the directors of both companies approved and entered into an agreement in principle for the acquisition of the business of National Steel Corporation (a wholly owned subsidiary of National Intergroup) by USS. The press release stated: "Under the agreement, United States Steel would acquire all of the steel-related businesses of National Steel Corporation, including its three steel plants, iron ore and coal operations." The price for these facilities was $575 million; in addition, USS would acquire part of the long-term debt associated with National Steel, which was over $300 million.

Both USS and National were anxious to have a quick decision from the Justice Department on the possibility of the acquisition. This was necessary, since National, due to the uncertainty, was suffering losses in the marketplace. On March 8, 1984, the Department of Justice indicated that it would oppose the merger unless some facilities were disposed of. After some negotiations the two companies agreed to terminate the arrangement, and National Intergroup was again firmly established in the steel business. With the failure of the merger to obtain approval.

Less than a month after the Justice Department's denial of the merger, National Intergroup made an announcement that it had agreed to sell 50 percent of its interest in its steel facilities to Nippon Kokan of Japan. The price was approximately $300 million. A new company would be formed with half of the directors from National Intergroup and half from Nippon Kokan. National hopes to gain in a number of ways, including a considerable infusion of cash and access to technology developed by Nippon Kokan, as well as the know-how in virtually every step of steel-mill operations. It is quite possible that the cash may be used to install a continuous-casting unit as well as an electrolytic galvanizing line at the Great Lakes plant.

The arrangement has been termed a joint venture and will have to receive government approval. Following as it does on the heels of Wheeling–Pittsburgh's liaison with Nisshin Steel, it indicates that there is a strong possibility that more international arrangements may be made in the future.

Notes

1. *1974 Annual Report*, National Steel Corporation, p. 3.
2. *1974 Annual Report*, National Steel Corporation, p. 4.
3. *1976 Annual Report*, National Steel Corporation, p. 4.
4. *1979 Annual Report*, National Steel Corporation, p. 15.
5. *1979 Annual Report*, National Steel Corporation, p. 4.
6. *1982 10-K Report*, National Steel Corporation, p. 1.
7. *Brochure of National Intergroup*, Pittsburgh, Pa., 1983, pp. 2, 3.
8. Ibid.

The Jones and Laughlin Steel Corporation

Jones and Laughlin (J&L) the oldest integrated steel company in the United States, ceased to be an independent when it was acquired by the conglomerate, LTV, in 1968. A decade later, J&L acquired Youngstown Sheet and Tube, which was on the verge of bankruptcy. The merger, allowed because Youngstown Sheet and Tube was judged to be a failing company, moved J&L into fourth place in the country in terms of raw-steel-producing capacity. Youngstown Sheet and Tube had already closed its Youngstown plant and was about to shut down the Indiana Harbor plant when the Justice Department permitted the acquisition. As a result, J&L has five steelmaking plants and one finishing facility. The five plants include:

1. The Aliquippa Works at Aliquippa, Pennsylvania, which is fully integrated with coke ovens, blast furnaces, basic-oxygen steelmaking facilities, and rolling mills for structurals, bars, tinplate, and seamless pipe.
2. The Indiana Harbor Works, the largest integrated plant in the corporation. At the present time, however, its coke ovens are not functioning and thus coke is brought in from other locations. This plant is mainly devoted to flat-rolled products, including sheets, tinplate, and galvanized steel. It also has seamless-pipe mills.
3. The plant in Cleveland, which is integrated except for coke ovens. The plant is completely devoted to the production of flat-rolled products.
4. The Pittsburgh Works, which now produces steel by the electric-furnace method and ships much of it to other plants, including Aliquippa and Cleveland.
5. The Midland, Pennsylvania, Works, acquired from Crucible Steel in 1983. This is an electric-furnace plant with two 175-ton furnaces used for the production of stainless steel.

In addition to steelmaking plants, J&L has a finishing facility, which was built at Hennepin, Illinois, in 1968; it consists of a cold reduction mill and a galvanizing line. The plant originally obtained its hot-rolled steel from the Cleveland Works and, after the merger with Youngstown Sheet and Tube, the supply came principally from the Indiana Harbor plant. Cleveland Works continues to supply small amounts. When Hennepin was first built the intention was to make it a fully integrated plant; however, with the decline in steel demand, this plan has since been abandoned. J&L also operates several plants for the production of cold-finished bars.

During the past fifteen years, J&L has realigned its facilities so that the plants are interrelated to a very high degree; in this respect, J&L is unlike the other major producers. In the early postwar period, J&L's integrated plants were reasonably self-sufficient insofar as each processed its own raw materials to produce finished products. Thus the plants at Pittsburgh, Aliquippa, and Cleveland were fully integrated and maintained a status somewhat independent of each other. The first major departure from this plant self-sufficiency came in the late 1950s when the coke ovens at Cleveland were abandoned, thus forcing the plant to seek its supply from other J&L facilities.

The construction of the Hennepin Works in Illinois as a finishing facility also meant that it would depend for its hot-rolled coils on other plants (as mentioned, first Cleveland and then Indiana Harbor).

Basic changes occurred at the Pittsburgh Works when the blast furnaces and sheet mills were abandoned and the open-hearth furnaces replaced with electrics. Pittsburgh produces more raw steel than it can finish; thus it supplies semifinished steel to Aliquippa, Cleveland, and Indiana Harbor.

The trend to interrelation between the plants was accentuated in 1981 when the hot-strip mill at Aliquippa was taken out of operation and hot-rolled bands made at Cleveland were sent to Aliquippa to be cold-reduced and then processed into tinplate.

The acquisition of Youngstown Sheet and Tube in 1978 brought with it a seamless-pipe mill in the Youngstown area. However, because the other facilities in the Youngstown plant were obsolete, they were closed down and the pipe mill was modernized with an investment of some $70 million. Steel for this mill in the form of tube rounds comes from Aliquippa, which also supplies tube rounds for the Indiana Harbor Works near Chicago. Because of the drastic drop in demand for seamless pipe, the seamless mills at Aliquippa have been closed down.

The Pittsburgh Works maintains its coke-oven batteries in operation to supply Cleveland and, to some extent, Indiana Harbor, where currently no coke is produced. It is interesting to note that none of the steel produced at Pittsburgh is used there. Its bar mills are supplied with steel from Aliquippa.

Aliquippa

Of the five coke-oven batteries that were operative in most of the postwar period, the Aliquippa plant now has two in operation. These batteries each have approximately five years of life left, after which time an investment of $50 to $60 million must be made for each battery to be rebuilt from the pad up. There are five blast furnaces standing; however, two of these have been abandoned and the plant is conceived of now as, at most, a two-blast-furnace operation with a third furnace on standby.

The Aliquippa plant was among the first in the United States to install the basic-oxygen steelmaking process. Two units, each with 85-ton capacity, were installed in the late 1950s. In 1968, they were replaced with three large 210-ton vessels, giving the plant a capacity of some 3.2 million tons of raw steel. Aliquippa is also equipped with a continuous casting unit which produces billets, blooms, and round sections. The finishing facilities consist of a 14-inch bar mill, which produces the renowned J&L junior structural beam, as well as bar products. The continuous hot-strip mill installed in 1957 has been shut down but not abandoned. It is doubtful however whether it will function again. There are two seamless pipe mills which cover a wide range of sizes. These are currently shut down and, unless business picks up appreciably, may not operate again. As previously indicated, hot-rolled bands are brought in from Cleveland to be cold-reduced on a five-stand cold-reduction mill and processed into tinplate. There is also a three-stand tandem mill for double-reduced tinplate, which when installed was the only one in operation in the world.

Cleveland Works

The Cleveland mill was purchased from Otis Steel Company in the early 1940s and, as indicated, was a fully integrated plant with coke ovens, blast furnaces, steelmaking equipment, and a strip mill for the production of sheets. In the early 1960s, J&L began revamping the plant, which was obsolete. The blast furnaces were small and steel was made by the open-hearth process; the hot-strip mill was also obsolete. One of the blast furnaces was replaced with a much larger unit capable of doubling the production. As the 1951 Annual Report stated:

> This furnace replaces a furnace of half its capacity. The increased blast furnace capacity permitted us to expand our open hearth steelmaking shop at Cleveland by two 210-ton furnaces.

In the late 1950s, as indicated, the coke ovens at Cleveland were abandoned and in 1963 the second small blast furnace was replaced by a new 2500-ton-per-day furnace. This was installed as part of a program at Cleveland Works to provide more pig iron for the basic oxygen converter. The latter was constructed at Cleveland Works in 1961 to replace the open hearths. Subsequently, two electric furnaces were added, giving the plant an annual capacity of 3.5 million tons.

Another major step in modernizing the plant was taken in 1964 when a new semicontinuous 80-inch hot-strip mill was installed. With the original facilities completely replaced, the plant currently is an efficient operating unit. However, it lacks a continuous caster and its cold-reduction mill (a four-stand facility) does not have the capability of producing a full line of sheets.

The market for Cleveland's product includes the automotive industry as well as appliances and welded pipe.

Any further investment in the plant would involve the installation of a new pickling line and an improvement to the cold-reduction mill.

Pittsburgh Works

The Pittsburgh Works was reduced in size when the open-hearth shop was replaced with electric furnaces. The original replacement plan called for a basic-oxygen shop; however, pollution problems led to the change in favor of electrics. Two 350-ton furnaces were installed and at full production these units can turn out 1.8 million tons of raw steel. With the installation of the electric furnaces, the blast furnaces were no longer needed and thus were abandoned. The coke ovens, however, which are in good repair, have been kept in operation to supply the Indiana Harbor and Cleveland Works. The sheet mill of prewar vintage was abandoned, and the finishing facilities are now limited to two bar mills. These are supplied with billets from Aliquippa and, as previously indicated, all of Pittsburgh Works's raw steel output is shipped in the form of slabs or blooms to other plants in the J&L organization.

The Pittsburgh coke ovens are very much needed to supply both Cleveland and Indiana Harbor and will continue to operate. The electric furnaces installed in 1979 replaced the open hearths and gave J&L a steelmaking ratio of 77 percent BOF and 23 percent electric.

Currently the Pittsburgh plant represents an unusual situation, since all of its production, in terms of coke and steel, is designated for other plants in the J&L complex.

Indiana Harbor

In 1978, with the merger with Youngstown Sheet and Tube Company, the Indiana Harbor Works of that organization became part of J&L with a plant located on Lake Michigan. This plant is fully integrated with coke ovens, blast furnaces, basic-oxygen steelmaking, and a modern 84-inch hot-strip mill as well as comparable cold-reduction facilities. In addition, there was a seamless pipe mill, so that Indiana Harbor was strictly a sheet and tube plant. Currently the plant has no coke-making operations, although there is an operable battery which has been closed down. With some investment, this can be reactivated. A second battery requires a pad-up rebuilt at a cost of $60 to $70 million, while a third battery that Youngstown Sheet and Tube contemplated building has never become a reality.

There are four blast furnaces standing; however, one has been abandoned so that only three are operable. Steel is produced from two 290-ton basic oxygen converters with an annual capacity of approximately 4.0 million tons.

A new continuous caster, capable of processing 3.0 million tons of slabs, has just been completed; it is probably the largest in the United States and was installed at a cost of $170 million. This will increase the yield from liquid steel to slabs by 12 percent to 14 percent and will allow the 84-inch hot strip mill to operate near its rated capacity.

Output from the plant consists of hot-rolled, cold-rolled, and galvanized sheets as well as tinplate. Currently the seamless tube mills are not functioning, but when demand picks up these will be put back in operation.

The Indiana Harbor Works, with the new continuous caster, revamped blast furnaces, oxygen steelmaking, and a modern strip mill, has the potential to produce quality-competitive steel. Its principal problem is the need to bring coke in from the outside. Currently, there are three sources: the Pittsburgh Works, Aliquippa, and coke made at Armco's Middletown Works from J&L's coal.

Future Development

In September of 1983, J&L announced that it would merge with Republic Steel. On February 15, 1984, the Justice Department stated that it would oppose this merger as presented. Immediately after this announcement, there was a vigorous reaction within the Reagan administration, as the Secretary of Commerce and the U.S. Trade Representative spoke out strongly against the Justice Department's decision. Negotiations followed, and on March 21, 1984, the Justice Department announced that an agreement had been reached which would permit the merger. Consequently, the future of J&L, which involves restructuring and rationalizing its facilities, will be determined by the course the merged companies follow. There are options open as a result of the merger that were not possible, had it not taken place.

Republic Steel Corporation

Republic was the third largest integrated steel company in the United States for several decades after its founding in 1930. However, with the merger of J&L and Youngstown Sheet and Tube in 1978, it dropped to fourth place in terms of steelmaking capacity. At the end of 1983, it operated six steelmaking plants, five of which were integrated and one, a specialty steel plant, which was supplied by electric furnaces.

The integrated plants that have blast furnaces and basic-oxygen steelmaking facilities include installations at:

1. Cleveland, Ohio, which was the largest of Republic's steelmaking operations and devoted primarily to the production of flat-rolled products.
2. Chicago, Illinois, which operates not only blast furnaces and bottom-blown oxygen converters but also electric furnaces. This plant's products include bars and seamless pipe.
3. Warren, Ohio, which in conjunction with some facilities at Youngstown produces primarily flat-rolled products of all types.
4. Gadsden, Alabama, which included the plant not only at this location but also the coke ovens at East Thomas, Alabama. The products here are sheets, strip, and plates.
5. Buffalo, New York, where the products are carbon and alloy steel bars, as well as semifinished steel.

The electric-furnace operation is in Canton-Masillon, Ohio, and has modern facilities (including an argon-oxygen decarburizer) that produce alloy and stainless steel, as well as specialty carbon steel products. The plant also has a continuous-casting operation.

During its corporate life dating to 1930, Republic has been first and foremost a steel producer with very little nonsteel interests. One significant venture into a nonsteel operation was announced in 1952 when Republic began commercial production of titanium and titanium alloys in several forms, including ingots, plates, sheets, and bars. Republic was optimistic about developments in this area as indicated in the Annual Report of 1952 which states: "While production is infinitesimal in comparison with steel production, demand for this new metal indicates that a substantial increase in production facilities may be soon required. Republic will continue to occupy a position of national importance in the production of titanium."[1]

The 1953 through 1957 Annual Reports list a variety of titanium products that the company produced including commercial pure titanium and titanium alloys, ingots, billets, hot-rolled and cold-finished bars, plates, sheets, and strip. The 1958 Annual Report makes no mention of titanium products since Republic discontinued this operation in that year.

In keeping with the rest of the industry, Republic decided to expand its steelmaking facilities to meet the growing demand for steel which hit its peak in 1973 and 1974. It also made provision to replace a number of facilities that were considered obsolete and inadequate. In this respect, it directed its capital expenditures in 1974 to the construction of oxygen steelmaking facilities in its Chicago area which replaced the open-hearth shop. It also replaced some coke-oven batteries at Cleveland and Warren, Ohio.

In 1975, a major expansion program was announced for the southern plant at Gadsden, Alabama. This expansion was calculated to add approximately 1.0 million tons at that location at a cost of $350 million. The announcement was made in May while the euphoria from 1973 and 1974 was still evident. However, there was caution expressed which indicated, "that the implementation of the plan was contingent on satisfactory profitability on a continuing basis. The company also cautioned that if environmental demands drained away too much of the available capital, there simply would not be enough left to implement the program. . . ."[2] The plan involved the construction of a large blast furnace, capable of producing 5000 tons of iron per day, as well as improvement to the plate mill. Unfortunately, with the decline in business, although much had been done in the way of engineering and site preparation, the program to construct the blast furnace was delayed and finally dropped.

In the late 1960s and 1970s, when other steel companies were diversifying, Republic cast its lot almost entirely with steel. This is its position in the 1980s as borne out by a statement in its 1982 10-K Report which stated, "For each of the years 1978 through 1982, steel products, which constitute the Registrant's predominant line of business, accounted for substantially all of the Registrant's sales, operating results and identifying assets."[3]

In the early 1980s, its capital investment program was directed primarily at the modernization of its seamless tube mill in Chicago, as well as the installation of improvements at the hot-strip mill at Gadsden. Unfortunately, however, with the decline in business, these improvement projects were cancelled. The only facility that was installed was a 2.0-million-ton continuous-caster for slabs at the Cleveland plant. At the close of 1982, Republic's steelmaking capacity at the six plants was estimated at approximately 11.3 million net tons.

Cleveland

The Cleveland Works is Republic's principal plant where steelmaking capacity from a basic-oxygen shop (with two 220-ton vessels) is approximately 2.8 million tons. The steelmaking facilities are supported by blast furnaces and coke ovens. There are four blast furnaces of which two are considered standby units so that the plant generally operates with two blast furnaces, each capable of producing 3000 tons of iron per day. There are relining schedules planned for these during the next two to three years, at which time considerable improvements will be made. Coke-oven batteries will also require a large investment during the next two to three years.

Before the installation of the new continuous caster, the Cleveland plant had far more finishing capacity than could be met by its steelmaking ability. The continuous hot-strip mill installed in the early 1970s can process 4.0 million tons of sheets if properly scheduled. Thus, the plant in times of peak demand could use slabs brought from other locations, either within or outside of Republic. However, the installation of the continuous caster (completed in late 1983) because of a higher yield will make more slabs available for the hot-strip mill and thus do much to balance melting and rolling facilities. It will also improve the sheet quality.

To keep the plant competitive, a substantial investment will be needed in the late 1980s for coke ovens and blast furnaces.

Warren

The Republic plant at Warren, Ohio, is tied to some extent to the Youngstown facilities, as indicated by the 1982 10-K Report, which describes the Mahoning Valley District as composed of plants at Youngstown, Warren, and Niles, Ohio.

Warren has one blast furnace which is the most productive unit in the corporation, having been substantially rebuilt in the last few years so that it is capable of producing 3400 tons of iron per day. This feeds two oxygen converters with a capacity of approximately 2.5 million tons. When these units were operated at near full capacity, the iron at Warren was supplemented from blast furnaces at Youngstown. However, since the steel industry operations have fallen drastically, the Youngstown iron was not needed and the two blast furnaces there, as well as the coke-oven batteries that support them, have been closed down and most probably will not be reactivated. Thus the steel output at Warren will be limited in the future by the iron supply, since its blast furnace can only furnish 1.3 million tons.

The high steel production figure at Warren was 1.9 million tons when iron from Youngstown was available.

The coke-oven battery at Warren (new in 1978) is adequate to supply the blast furnace. Warren's finished products consist entirely of sheets, including hot-rolled, cold-rolled, and galvanized, as well as terne plate. At high rates of production, the iron and steelmaking facilities at the plant are inadequate and, with the Youngstown facilities abandoned, they will remain so.

Chicago

The Republic plant in Chicago includes a new coke-oven battery installed in 1982 at a cost of $185 million. It supports the one blast furnace at that location, which in turn makes iron for the Q-BOP with its two 225-ton vessels. Because iron is limited, the Q-BOP is rated at 1.2 million tons. Steel production is supplemented by three 200-ton electric furnaces, capable of producing 600,000 to 700,000 tons of steel. The output of the mill in terms of products consists principally of bars with some rod production. As is the case with Warren, the Chicago plant output is limited by iron supply.

Buffalo

The Buffalo plant at which operations were suspended in 1982 because of poor business conditions has been permanently shut down as of January 1984. The plant consisted of four coke-oven batteries owned and operated jointly with the Hanna Furnace Company, a division of National Steel. These provided coke for Hanna's three blast furnaces, which have now been torn down. With the closure of the Republic plant, there is no need for the coke ovens. The Republic plant had two blast furnaces (one of which had been recently relined), two oxygen converters with a 1.0 million tons capacity, and several bar mills. The closure of the plant permanently eliminated 1.0 million tons from the steelmaking capacity of the United States.

Gadsden

The Gadsden plant has furnished steel markets in the South for over fifty years. It was acquired by Republic in 1937 and since that time has been a significant producer of sheets and plates. It has two batteries of coke ovens, which are supplemented with a third battery located at East Thomas, Alabama. These support two blast furnaces, one of intermediate size and the other quite small. The two basic-oxygen converters are rated at 1.5 million tons of annual steelmaking capacity. The output is rolled into plates and sheets, some of which are galvanized. The coke-oven battery at East Thomas has been shut down and it is unlikely that it will be reactivated. Further, one of the batteries at Gadsden is in poor condition and will require a major rebuild from the pad up if it is to continue to operate beyond 1986. The second battery is in reasonably good condition and could continue to operate for several years without any major investment.

The Number 2 battery, which is approaching the end of its life span, has been the subject of an EPA complaint against Republic for extensive emissions. The consent order entered into required Republic to proceed with an air-pollution control program at an estimated cost somewhere in excess of $500,000. It also levies daily fines for future violations and unquestionably will speed the shutdown of this battery. With one coke-oven battery and one blast furnace out of operation, the plant will be considerably reduced in terms of its steelmaking capacity. The remaining blast furnace can produce approximately 900,000 tons of iron, which is not enough to keep the oxygen converters operating at capacity. In the future the maximum steel production with one blast furnace providing iron would be slightly more than 1.0 million tons. This could be supplemented with slabs from an outside source.

The merger between Republic and J&L, which was approved by the Justice Department on March 21, 1984, stipulates that the Gadsden plant should be sold. Once this is accomplished, it no longer will represent a problem for the merged company.

Canton-Massillon

The Canton-Massillon facilities are based on electric-furnace operations and produce alloy as well as stainless steel. As a result of the merger decision, the new combination will have to sell Massillon's cold-rolled finishing facilities for stainless steel. However, an agreement was made to provide hot-rolled stainless steel bands for a ten-year period to the purchaser of these facilities.

Republic's performance in terms of earnings has been quite satisfactory through 1981, when it earned $190 million. Part of this, however, came from the sale of coal reserves in Alabama to the Gulf Oil Corporation for approximately $72 million. In 1982, along with the rest of the industry, Republic incurred a substantial loss of $239 million. Losses, amounting to $326 million, continued in 1983; however, a significant amount of this was due to the writeoff of the Buffalo plant, which amounted to $194 million.

The closure of the Buffalo plant represented a significant step in the restructuring of the company. However, although others could well have been taken, the proposed merger with the J&L Division of LTV delayed further action.

Now that the merger has been consummated, the best of Republic's facilities will be combined with the best of J&L's and should result in a strong steel company. Fortunately, Republic brings an abundance of raw materials to the merger, since it has interests in iron ore reserves and pellet plants in the United States and Canada, as well as coal mines in several eastern states.

Notes

1. *1952 Annual Report,* Republic Steel Corporation, p. 12.
2. *1975 Annual Report,* Republic Steel Corporation, p. 6.
3. *1982 10-K Report,* Republic Steel Corporation, p. 2.

10 Republic-J&L Merger

In September of 1983, J&L announced that it planned to merge with Republic Steel to form a new company to be known as LTV Steel. The new entity would have an initial nominal raw steelmaking capacity of some 23.0 million tons annually. With this tonnage, it will be the second largest steel company in the country, moving ahead of Bethlehem Steel Corporation, which held that position for sixty years. On February 15, 1984, the Justice Department announced that it would challenge the merger unless some modifications were made. Negotiations over a period of a month resulted in a proposal which was acceptable to the Justice Department, and the merger was allowed.

The two companies agreed to sell the Republic plant at Gadsden, Alabama, as well as Republic's cold-rolling facilities for stainless steel at Massillon, Ohio. Further, they agreed to provide hot-rolled stainless steel bands to the purchaser of the Massillon facilities for a period of ten years. Under the terms of the merger, LTV and Republic must complete the sale of both plants within six months or name a trustee to sell the facilities.

The restructuring of the corporation and the rationalization of facilities offer a number of options. Both companies have operating facilities in the Chicago, Cleveland, and Youngstown areas. The plants in Cleveland are adjacent to each other, and both produce flat-rolled products. The two plants have modern hot-strip mills, although Republic's was installed more recently and is a wider mill. Further, Republic's cold-reduction mill, installed at the same time as the hot-strip mill, is an inherently more efficient and competitive mill than the one at J&L's plant. Republic has installed a new continuous-casting machine to process slabs for the hot-strip mill. Without doubt, some rationalization of the facilities in both plants can be arranged, so that operations will be combined, costs reduced, and efficiency increased.

With the sale of the Gadsden plant, the company will no longer produce steel in the Southeast. However, the elimination of Gadsden relieves the new combine of the need to invest large sums of money for modernization, since the blast furnaces and coke ovens need considerable attention.

There will undoubtedly be a number of changes in the size and composition of J&L's plants. The Pittsburgh plant will continue to operate its coke ovens to supply Cleveland and Indiana Harbor. This is a necessity, for even with the new combination the J&L plants are short of coke. The electric furnaces at Pittsburgh represent a recent large investment, and it is difficult to see how these units can be discarded.

The Aliquippa plant, because of the collapse of the seamless pipe and tube market, has operated at about 25 percent of its raw steelmaking capacity during the past year. Its seamless tube mills, which are not competitive with the new facilities installed in the United States and abroad, will most probably not operate in the future unless demand increases to the 1981 level, and this is highly unlikely.

Aliquippa, with its continuous casting equipment for the production of tube rounds, should function to supply the seamless mills at Youngstown and Indiana Harbor. The plant will also continue to receive hot-rolled bands to be converted into tinplate.

It seems virtually certain that Aliquippa will never produce at its rated raw-steel capacity of 3.0 million tons. Under these circumstances, it is entirely possible that the blast furnaces and BOF vessels could be replaced by electric furnaces in the future.

The Indiana Harbor plant of J&L, with its continuous caster, will continue as one of the mainstays of the new corporation.

In terms of raw materials, both companies have adequate supplies. In respect to iron ore, J&L has a 35 percent interest in both the Empire Mining and Erie Mining Companies. The latter was obained when J&L acquired Youngstown Sheet and Tube. J&L also has a 12 percent interest in the Tilden Mine.

Republic has a 50 percent interest in the Reserve Mining Company as well as a 16 percent interest in the Hibbing Taconite Company. Pellet plants are in operation at all of these locations.

Both Republic and J&L have ore interests in Canada as partners in the Iron Ore Company of Canada and Wabush Mines. With respect to coal, both companies have mines in several locations in the Appalachian area.

It will take some time to coordinate the plants; however, the facilities of the combined company will be restructured, resulting in a stronger, more efficient operation with higher productivity and lower costs.

Two top-level executives of the new combine affirmed this in a statement which noted that steel is "an intensively competitive worldwide business and the steps we have taken to date are not enough."

11 Weirton Steel

In March 1982, National Steel Corporation made the following announcement:

> As part of its ongoing effort to direct capital funds to areas of higher return, it had decided to substantially limit its future capital investments in the Weirton Steel Division, Weirton, West Virginia.[1]

In order to maintain the plant in existence rather than ultimately shut it down—which would have happened if no capital expenditures were made—National offered to sell the plant to the employees. This decision was reached after considering a number of possibilities. The result was expressed by Howard M. Love, National's Chairman:

> We believe employee ownership of the Weirton Division is a viable option in meeting our responsibility to our stockholders, Weirton Steel employees, Weirton and surrounding communities and our customers. Employee ownership appears to be econonically feasible at Weirton and is preferable to any other alternative.[2]

During the following year, a number of studies were undertaken concerning the feasibility of operating the Weirton plant as an independent unit. The principal study was made by McKinsey and Company, Inc.; its July 1982 report concluded that with a number of changes the plant could be a viable economic entity. A substantial reduction in production costs was required and much of this was to be obtained through a 32 percent wage reduction for all employees, including management. This meant that many of the employees who were members of an independent union and receiving $24.90 an hour had to sustain a reduction of $7.90 per hour. Other employees with different incomes also were required to take a 32 percent cut.

The proposal and the suggestions were discussed and debated at great length among the employees at Weirton and, in September 1983, a vote was taken by which they agreed to buy the plant and take a reduction in their income. This was an unprecedented action for Weirton employees, who enjoyed a wage rate several dollars per hour above that of the United Steelworkers in other plants. The decision was reached since it was a matter of survival; had the employees voted against acquiring the Weirton plant,

National Steel would have materially reduced its size and most probably closed it down within a few years. National's Chairman made this quite clear in an announcement that stated:

> National has embarked on a program of "downsizing" its steel facilities to whatever size is required to meet market demand and earn a return for its stockholders.[3]

This statement was taken seriously by the people at Weirton as the prospectus issued in connection with the acquisition stated:

> National has also taken under consideration the closing of one of the two remaining blast furnaces. Furthermore, it is possible that other National planning alternatives for Weirton could include a shutdown of Weirton's primary end, reducing the mill to a finishing operation. It is also possible, for the long-term, that National could completely terminate the operations of the division.[4]

The acquisition price for the plant and all its facilities in terms of cash was $74,680,861, in addition to which there were two long-term notes, one due in 1993 for $47.2 million and a second due in 1998 for $72.0 million. This was a reasonable price when one considers that the plant was valued at some $314.4 million.

One might ask the question: Why did National divest itself of this relatively modern plant if it thought the employees could operate it profitably? In response, National stated that its decision to ask the Weirton management and its unions to consider employee ownership was not the result of economic conditions but corresponded to the overall corporate plan that had been developed to provide a substantially improved return to the stockholders of National Steel on a long-term basis. Further, National stated:

> Even though Weirton is equipped with modern basic-oxygen furnaces, continuous slab casting, and the most advanced computer-controlled cold rolling mill in the United States, its capital requirements during the next few years will be substantial . . . and
>
> It is not economically feasible for National Steel to commit such large amounts of capital at Weirton since funds would have to be diverted from other projects throughout the Corporation which would have the potential for substantially higher return than at Weirton.[5]

On January 11, 1984, the final transfer of the plant to the new Weirton Steel Corporation was made. Thus a new independent integrated steel operation came into being. The company is owned by the employees through an Employee Stock Ownership Plan (ESOP), and although the employees have stock in the company, they signed a collective bargaining agreement with the newly constituted management.

It provided for a 32 percent cut in wages and a wage freeze for six years. The contract is to run for three years, at which time it is reopenable for noneconomic matters. Wages and other economic questions will not be renegotiated for six years. There was also a no-strike agreement and the elimination of cost-of-living adjustments. The incentive for this was a promise of a share in the profits once the company's net worth exceeded $100 million. It is interesting to note that it would have been impossible for National to have obtained such concessions if it maintained ownership and operation of the company. In a statement entitled "Weirton Joint Study Committee, Inc." designed to be a disclosure document on the acquisition of the assets of the Weirton Steel Divison of National Steel Corporation and the establishment of the Weirton Steel Corporation, the following comment relating to the concessions was made:

> If the closing does not take place for any reason, such collective bargaining agreements and amendments to existing agreements with National will be of no effect. The employees of the division will not be obligated to give any concessions to National.[6]

The viability of Weirton as an independent company is closely related to the tinplate market, since the Weirton plant is very heavily dependent on tinplate. During the last six years, from 1977 to 1982, although tin-mill products output at Weirton had declined as they have throughout the steel industry, Weirton's market share remained steady. In 1977, shipments of tin-mill products for the industry were 6,383,000 tons, of which Weirton shipped 1,162,000, or 18.2 percent. In 1980 and 1981, the figure rose to 20 percent, and in 1982, when industry production was 4,322,000 tons, Weirton shipped 817,000 tons, representing 18.9 percent. In addition to tinplate, hot- and cold-rolled sheets as well as galvanized materials, will continue to be produced. In this regard, Weirton has a five-year commitment to supply 180,000 tons of hot-rolled bands per year to National Intergroup's Midwest plant.

Capital requirements for the new corporation have been put at $80 million to $100 million a year for the next five years. It is doubtful, however, that this much can be spent (certainly not in the next two or three years) since it will be necessary for Weirton to conserve its cash and operate as prudently as possible.

Generally speaking, the Weirton plant is a competitive unit with a number of modern facilities, as well as some that should be updated and improved. However, the company is fighting for survival and can not afford the large capital investment required to do this. For the next two to three years, it must make do with what it has until financial conditions improve.

At present, there are no coke ovens operating or operable at Weirton. All have been shut down and a new battery, which would be necessary, will require a large investment of approximately $200 million. This is out of the question at the present time. Coke is currently obtained by purchase from the Bethlehem plant at Lackawanna and the Shenango ovens on Neville Island, near Pittsburgh. There are also other possibilities, including imports from Japan and Western Europe.

The blast-furnace situation has improved considerably. There are four operable furnaces. Presently, three (Numbers 2, 3, and 4) are in blast. Number 2 was placed in operation in February 1984, and after a reline Number 3 was placed in operation in late March. At that time, Number 1 was taken off to be relined so that the plant will operate three furnaces during the remainder of the year. Enough iron will be produced to feed the basic-oxygen converters, although they will not be able to attain their capacity of 3.5 million tons. Actual production for 1984 will be between 2.1 million and 2.3 million tons.

Weirton has a four-strand continuous caster for slabs, which was one of the first installed in the United States some fifteen years ago. This will require some investment for modifications during the current year. The hot-strip mill is pre-World War II vintage. However, it has been improved and updated from time to time and functions adequately. One outstanding facility at the Weirton plant is a five-stand cold-reduction mill, the most recent to be installed in the United States and, consequently, the most modern. There are also a number of tinning lines, as well as lines for the production of tin-free steel.

The new Weirton steel company can succeed, since it has good facilities, some of which are excellent; access to raw materials at good prices; a reduced cost structure; and a reputation for making an excellent tinplate product. Its raw steel output will most likely remain at 2.5 million tons a year for the remainder of the decade although the market for its principal product is declining at a moderate rate of 1 to 2 percent a year. However, there is still a significant market for tinplate and the decline is not serious enough to affect the viability of the Weirton plant.

Another factor in its favor—at least at the beginning of the operation—is the determination of its employees, who saved the plant from extinction, to succeed.

Notes

1. *Press Release*, National Steel Corporation, March 2, 1982.
2. Ibid.
3. *Press Release*, National Steel Corporation, March 2, 1982.

4. Weirton Joint Study Committee, Inc., Weirton Steel Corporation. *Document of Disclosure*, August 29, 1983, p. 27

5. *Press Release*, National Steel Corporation, March 2, 1982.

6. Weirton Joint Study Committee, Inc. Weirton Steel Corporation. *Document of Disclosure,* August 29, 1983, p. 43.

12 Wheeling-Pittsburgh Steel Corporation

Wheeling-Pittsburgh Steel Corporation was formed in the closing months of 1968 as a result of a merger between Pittsburgh Steel Corporation and Wheeling Steel Corporation. As a consequence, the new entity had major integrated plants in two locations. Wheeling Steel had its principal plant at East Steubenville, West Virginia, near Wheeling, with some finishing facilities at Yorkville, Ohio. Pittsburgh had its integrated steelmaking operation at Monessen, Pennsylvania (not too far from Pittsburgh), with finishing facilities at nearby Allenport, Pennsylvania.

In 1982 and 1983, the company undertook a major capital investment program that involved expenditures for two continuous casting units, one at Steubenville for slabs for the hot-strip mill and the other at Monessen for blooms for the newly constructed rail and structural mill, as well as for the seamless pipe mill. Both continuous casters went into operation in late 1983 with minimum startup problems.

The rail mill has operated since 1983 and has a capacity to produce 400,000 tons of quality rails in lengths up to 82 feet. Currently, this mill is one of three in the United States producing rails with the closure of USS's rail mill at Gary, Indiana, in April 1984. The other two are located at Bethlehem's plant at Steelton, Pennsylvania, and CF&I Steel's plant at Pueblo, Colorado. It appears from an examination of the demand for and production of rails during the last ten years that the three mills that will remain have an adequate capacity to meet the requirements of the railroads. Since 1973, railroad demand for rails has averaged about 1.0 million tons a year, with the high point of 1.3 million tons reached in 1976 and the lowest level of slightly under 500,000 tons in 1982. Table 12-1 gives rail shipments for the past decade.

The three mills that will be left after April 1984 have an aggregate capacity in excess of 1.1 million tons, which on an average basis is adequate to take care of rail demand if it continues as it has for the last ten years. It is more than likely that these mills will operate at almost full capacity in the years ahead.

The two continuous casters at Wheeling-Pittsburgh have now made it possible for the company to process 70 percent of its raw steel through these units, a very high rate by comparison with the average in the United States.

The company produces a variety of products considering its size. It has a total capacity of 3.5 to 3.8 million tons from which it produces rails, struc-

Table 12-1
Rail Shipments: 1973-1983
(thousands of tons)

Year	*Shipments*
1973	916
1974	924
1975	1179
1976	1303
1977	1109
1978	946
1979	1122
1980	1085
1981	907
1982	498
1983	610

turals, seamless pipe, hot- and cold-rolled sheets, tinplate, and galvanized sheets. The two plants at Monessen and Steubenville have a significant number of modern, competitive facilities. However, there are also facilities that leave something to be desired.

Monessen

The Monessen plant has a capacity to produce 1.3 to 1.4 million tons of raw steel when it is in full operation. It has three blast furnaces standing, two of which are small and one medium size as American blast furnaces go. The medium size furnace (Number 3) is operating, while one of the two smaller furnaces has been abandoned and the other kept in readiness as business requires or when Number 3 is taken down for relining. The furnaces are supported by two operable coke-oven batteries that have been rebuilt from the pad up in 1979 and 1980. The third battery has been down idle cold and most probably will not function again. Steelmaking facilities consist of two 200-ton basic-oxygen converters.

The principal finishing facilities are located at Monessen and at Allenport, a short distance away. The continuous caster is at Monessen, along with a newly constructed rail and structural mill. At Allenport there is a 66-inch hot-strip mill along with equally wide cold reduction facilities and two seamless-pipe mills. It is quite likely that slabs for the Allenport strip mill may be supplied from the continuous slab caster at Steubenville.

Steubenville

The plant at Steubenville has four coke-oven batteries and five blast furnaces. The coke-oven batteries are all operable. However, it is quite likely that two of the blast furnaces will not operate in the future, and the plant will be a three blast-furnace operation. These units provide enough iron for the basic-oxygen converter to produce 2.2 to 2.4 million tons of steel, virtually all of which will be continuously cast into slabs and rolled on a modern hot-strip mill. The finished product from Steubenville consists of hot-rolled and cold-rolled sheets, as well as galvanized sheets and tinplate.

The slab caster will improve the quality of the products and increase the yield so that operations will be more efficient and, with the increase in demand for flat-rolled products, should be profitable.

In 1983, Wheeling-Pittsburgh approached Siderbras, the Brazilian steel-holding company, to purchase several hundred thousand tons of slabs on an annual basis provided the Brazilians would make an investment at Wheeling. Slabs were to be supplemental rather than a replacement of those made domestically. Negotiations came to nought.

On February 7, 1984, Wheeling-Pittsburgh made a dramatic announcement that an agreement had been reached with Nisshin Steel of Japan, the smallest integrated company in that country. Each will purchase the others stock and engage in a joint venture on a coating line that will require a $40 million investment. Nisshin has developed an expertise in coating steel and this technology will be available for the new facility.

Wheeling-Pittsburgh has obtained an infusion of capital as well as a partner in the construction of a very significant facility for steel sheet products. Although it is too early to pass judgment, the joint venture and quasimerger appears to be decidedly beneficial for Wheeling-Pittsburgh. This, coupled with its new continuous casters and new rail and structural mills, should provide the company with a basis for profitable operations in the future.

The Other Integrated Companies

McLouth Steel Corporation

McLouth Steel is a relative newcomer to the industry, having been established in its present state after World War II. The plant, located at Trenton, Michigan, near Detroit, was financed in part by General Motors to provide it with steel at a time when steel was in tight supply. The facilities, at the outset, were not completely integrated since coke had to be purchased. In subsequent years, coke ovens in various locations were acquired. The company had a varied financial history, fluctuating between considerable losses and good profits.

In December 1981, due to poor business conditions—particularly in the automotive industry which suffered two bad consecutive years—McLouth filed for Chapter 11 and it appeared that the plant would be liquidated. One result of the Chapter 11 filing was the relinquishing of the coke oven property at Portsmouth, Ohio. Fortunately, in November 1982, before the liquidation process began, the company was purchased by Cyrus Tang and put back in operation.

The facilities consist of two identical blast furnaces, each capable of producing 2700 tons of iron per day. Only one of these is operating at present, and it will be necessary to reline the second if business conditions improve to the point where it is needed. The steelmaking sector consists of five 120-ton BOF vessels and two 200-ton electric furnaces. McLouth was the first company in the United States to install a basic-oxygen converter and the third in the world; this was in 1954. In addition, McLouth was also the first major integrated company to adopt continuous casting for its entire steel production. There were severe problems at first but these were finally overcome. Finishing facilities consist of a 60-inch-wide hot-strip mill and comparable cold-reduction mill, as well as a line for galvanized sheets.

McLouth's principal market is the automotive industry, which takes a major portion of its output.

At the present time, McLouth buys all of its coke from Elk River Resources. This is just enough to operate one blast furnace. Should it be necessary to bring on a second furnace, an additional 50,000 tons of coke per month would be needed. In respect to iron ore, McLouth has no holdings of its own and must buy its entire requirement. This is obtained from various sources such as Empire, Tilden, and Erie mines and pellet plants. The electric furnaces are in operation and were low-cost producers with the price of scrap at $60 a ton in early 1983. However, recently it has risen to about $100, which caused a corresponding increase in electric-furnace steel costs.

Investments and improvements currently under consideration are directed principally to cost reduction. On the continuous caster, which processes all of McLouth's steel, various width molds are being installed; on the hot-strip mill, new gauge and width controls as well as a new computer have been added. At the cold mill, gauge controls have been improved. Further, in order to reduce energy, some of the batch annealing furnaces will be replaced, thereby cutting the BTU use per ton from 1.7 million to 700,000. There is also provision for desulfurization of iron outside of the blast furnace, which will allow for greater production and a better coke rate. As is evident, these improvements are designed to reduce costs and increase yield.

With two blast furnaces and two electric furnaces in operation, McLouth can produce 2.6 million to 2.7 million tons of raw steel. However, under the present circumstances with one blast furnace and two electrics in operation, raw steel output is approximately 1.5 million tons.

In addition to the above-mentioned improvements, very little change is expected at McLouth during the next two years. The company will continue to produce flat-rolled products, which will be primarily for the automative industry and steel service centers. It currently has a joint venture with a steel service center to install a galvanizing line for heavy sheet (both hot- and cold-rolled).

Rouge Steel Corporation

Rouge Steel, formerly known as the steel division of Ford Motor Company, now a wholly owned subsidiary of Ford Motor Company, has been in operation for more than sixty years. The name was changed in 1982 and its status shifted from a division to a subsidiary, thus presenting some advantages in terms of possibilities of selling the facilities. In late 1982, negotiations were undertaken with Nippon Kokan of Japan in the hope that Rouge could be sold either in whole or in part (75 percent) to Nippon Kokan. These negotiations broken down in mid-1983 and Rouge was then faced with the decision to close the plant or invest in much-needed equipment.

The company was organized by the United Autoworkers rather than the Steelworkers and employment costs were some $5.00 an hour above the Steelworkers' level. Further, the plant had shortcomings insofar as it did not have a continuous caster and its cold-reduction mill was not adequate to produce the high-quality sheets required for exposed parts of automobiles. The labor contract was renegotiated to bring wages much more in line with those paid by the other steel companies. As a result, a pledge was given at that time to continue the plant in operation.

The plant at Dearborn, Michigan, is fully integrated and produces predominantly flat-rolled products, a large portion of which in some years (as much as 65 percent) were taken by the Ford Motor Company. In recent years, this dropped to as low as 35 percent.

At the close of 1983, the coke-oven segment of the plant consisted of three batteries of ovens, one of which has been shut down so that only two remain operating. Whether the unit that has been closed will be revived is a serious question to be settled in the next year or more. To provide adequate coke for the blast furnaces, an arrangement was made to obtain the full output of the plant in Portsmouth, Ohio, which formerly belonged to Cyclops. Some 400,000 tons are available to Rouge from this source, under a two-and-one-half year contract with an option to renew for two-and-one-half additional years.

There are three blast furnaces; two are quite small because they were built in the 1920s when the plant was constructed. The third furnace, built after World War II, is a medium-size American blast furnace and has within the last year been completely rebuilt with a new furnace stack, two new stoves, and a new skip hoist. With these improvements, the furnace will be capable of producing 3500 tons of iron per day. The smaller furnaces have a combined output of approximately 3500 tons a day. One of these will have to be substantially rebuilt in the next year or two.

Steelmaking is carried on in a basic-oxygen shop with two 240-ton furnaces and an electric-furnace shop with two 200-ton furnaces. Total capacity from both units is 3.6 million tons of raw steel. The oxygen furnaces were installed in 1964 and the electrics in 1976. The hot-strip mill is the most recent unit installed in the United States, having been put in operation in 1974. It is 68 inches wide and has a capacity to produce approximately 3.0 million tons of hot-rolled bands. As indicated, as of 1983 the cold-reduction mill left much to be desired. However, in the past year it has been renovated with better gauge controls as well as computer controls, making it possible to produce cold-rolled sheets for exposed automobile parts. The company has recently announced its intention of installing a new continuous casting machine with a capacity to process 1.8 million tons of steel. This will improve the quality of its sheets and reduce costs by more than $30 a ton. Mitsubishi of Japan has the contract and very attractive financing has been provided, which allows Rouge maxium flexibility.

Since the automotive industry is turning more and more to galvanized sheets for corrosion-resistant bodies, Rouge is contemplating the installation of a galvanizing line within the next two years.

The restructured company with a new continuous caster and improved cold-reduction mill (and possibly a galvanizing line) should make Rouge more competitive, particularly since its labor cost is now in line with that of the industry. One distinct advantage that the company's improved facilities have is the sale of at least 40 percent of its output to the Ford Motor Company.

Rouge Steel has an iron-ore supply represented by its 85 percent interest in the Evelyth Taconite Company. Coal is obtained from mines in Kentucky.

Sharon Steel Corporation

Sharon Steel Corporation became a part of a conglomerate when NVF, a producer of plastics, took it over in 1968. It was and remains the largest division of NVF. Sharon has one integrated steel plant located on the borderline of Pennsylvania and Ohio; formerly this plant belonged to USS and was known as the Farrell Works until 1945, when it was sold and became the major part of Sharon Steel Corporation.

As of 1984 the plant has the capacity to produce 2.0 million tons of raw steel annually. It has two operable blast furnaces capable of producing 1.3 million tons of iron per year and a BOF shop with two 150-ton vessels that can produce 1.5 million tons of steel annually. These two conventional BOF vessels replaced two Kaldo oxygen vessels, which were found to be less economical than the conventional units. In addition to the BOFs there are two electric furnaces each with 130-ton heat capacities. Annual capacity from these units in full operation is approximately 500,000 tons.

The plant is almost exclusively a flat products facility, with a slab and blooming mill, a hot-strip mill and cold-reduction mill, as well as a hot-dip and an electrolytic galvanizing line.

Sharon Steel has no coke supply and must depend entirely on purchases. During the past years it has obtained coke from Shenango, Youngstown Sheet and Tube, Republic, Armco, Jewel, and a number of others. With no financial interest in any one coke operation, Sharon depends on a variety of sources. For its ore, it has a share in the Tilden Mines.

The company's sales consist of a large number of relatively small-size orders and a wide variety of products. These include: hot- and cold-rolled sheets, as well as hot-dipped and electrolytic galvanized sheets. In addition, there are high carbon strip as well as forging-quality steels. Because of this, Sharon does not find it economical to invest in a continuous caster and will continue to operate on an ingot basis.

There have been discussions within the corporation about its future which involve shutting down one blast furnace and adding another electric furnace with a 200-ton per heat capacity. This would result in a reduction in raw steel capability to about 1.8 million tons annually; it would also lessen the dependence of Sharon on outside coke. With two blast furnaces Sharon frequently finds it necessary to buy slabs when one furnace is down; Sharon will continue to be a potential slab purchaser under these circumstances.

CF&I Steel Corporation

CF&I is an example of a company whose structure has changed radically over the last twenty-five years. In 1957, it operated four plants producing steel, one independent blast furnace, and three wire drawing plants. The principal plants were:

1. The Pueblo Works at Pueblo, Colorado, a completely integrated unit.
2. The Wickwire Spencer Steel Company near Buffalo, which was an integrated operation. This was closed down in 1963 since it was judged to be an obsolete, high-cost plant.
3. The Claymont plant at Claymont, Delaware, which was a semi-integrated operation based on open-hearth furnaces, charged with cold metal, and a plate mill. This was sold to Phoenix Steel in 1960 for $15 million.
4. E. & G. Brooke plant at Birdsboro, Pennsylvania, which consisted of a blast furnace, producing foundry and malleable pig iron. It was closed in 1962, having been judged to be obsolete and high cost.
5. The Roebling plant at Trenton, New Jersey, based on a cold-metal open-hearth shop which produced steel for wire rods and wire. The open-hearth shop was replaced by electric furnaces in 1964. However, the plant was uneconomical and was closed in 1974. Much of it was sold piecemeal in the subsequent year or two.

Thus, since the mid-1970s, CF&I has consisted of the integrated plant at Pueblo, Colorado. In 1983 it had a capacity to produce approximately 2.0 million tons of raw steel annually.

In 1957, the Pueblo plant was an open-hearth operation, as far as steelmaking was concerned, with approximately 1.8 million tons of capacity. In 1961, the open hearths were replaced with two 120-ton basic-oxygen steel converters, and in 1973 and 1976, two electric furnaces were added, each with 150-ton capacity. The basic-oxygen converter shop had a 1.3-million-ton capacity, while the electric furnaces could produce approximately 650,000 tons annually. In 1972 and 1973, over 100 coke ovens were rebuilt.

As of 1983, the plant had three coke-oven batteries and four blast furnaces providing iron for the BOF. In terms of finishing facilities, CF&I had a rail mill, which was updated and modernized in 1979 so that it could roll rails 82 feet long. It also had a seamless-pipe mill, structural mill, several bar mills, and a rod mill. In 1976, a continuous caster was installed for billets.

The booming oil-country tubular-goods market, which took huge tonnages in 1980 and particularly in 1981, inspired the CF&I management to build an additional seamless pipe mill, with about 160,000-ton-capacity, along with a continuous caster for round sections to feed the mill. The investment was calculated to be approximately $175 million. Work was started on the project, and the continuous caster was finished. The new caster is used to feed the existing seamless pipe mill. For when the oil-country tubular goods market collapsed in 1982, the new pipe mill was placed on hold. In 1983, it was abandoned as the oil-country tubular-goods market hit an all-time low. The total shipments for that year were 670,000 tons as compared to 4.3 million in 1981.

In 1982, with business at a low ebb, the coke ovens, blast furnaces, and oxygen steelmaking converters were all shut down so that the only steel made at Pueblo was in the electric furnaces. This was enough to feed the plant, since the demand for its major products, rails and oil-country goods, fell off sharply. Faced by these conditions and severe financial losses, the Crane Company, which acquired CF&I in 1969, was anxious to sell the plant. It was offered to a number of steel companies, both in the United States and abroad. At one time, serious negotiations were held with USS; however, they ultimately came to nought.

In December 1983, Crane Company made an announcement to the effect that the coke ovens, blast furnaces, and basic-oxygen converters at the Pueblo plant would be permanently closed. The 8-K form report dated December 27, 1983, and presented to the Securities and Exchange Commission stated:

> During December, 1983, CF&I Steel Corporation entered into the following transactions, the net result of which is to restructure its operations to reduce future costs and to improve its financial position through the sale of properties and elimination of debt. CF&I will hereafter concentrate steel production in its electric arc furnaces, utilizing its continuous caster and finishing mills for the production of rails, tubular goods, wire products, and structurals. The restructured format of operation has been in effect since mid-1982 when the decision to suspend coal mining and to bank the "hot end" (blast furnaces, basic oxygen furnaces and coke oven batteries) was made.
>
> On December 27, 1983, CF&I announced the permanent shutdown of its four blast furnaces, basic oxygen furnace, and three rolling mills in Pueblo, Colorado, and the commencement of an orderly shutdown of its coke ovens . . .[1]

The coke ovens were shut down in an orderly fashion, so that the damage to refractories would be minimal in the event someone would wish to purchase and reactivate them. Henceforth, CF&I's raw steelmaking capacity will be provided by the electric furnaces. Currently, the two electric furnaces are capable of producing some 650,000 tons per year; however, they can be improved and production increased by approximately 50,000 tons to 700,000 tons.

The rail mill, pipe mill, and rod and bar mills are still active. The rail mill has a capacity of 400,000 to 450,000 tons; the pipe mill 150,000 tons; and the other products, including bar, wire rods, and wire, should absorb an additional 150,000 tons. As demand for any of these products rises above the capability of the electric furnaces to produce steel, CF&I will purchase billets.

CF&I, having abandoned 1.3 million tons of capacity, will no longer function as an integrated mill but as an electric-furnace operation with a variety of products. It can, by no stretch of the imagination, be called a minimill, even though its tonnage is limited since its products are definitely not the type produced by minimills. In terms of corporate structure, the company may be spun off by Crane, with the current management retained to operate the remaining facilities and stock ownership distributed among the Crane shareholders. Crane may also sell CF&I.

Interlake Incorporated

Interlake Incorporated is one of the most diversified integrated steel producers in the U.S. industry. The present corporation was formed in 1964 as a result of a merger between Interlake Iron and Acme Steel Company. At that time Interlake Iron produced ferroalloys in addition to pig iron, while Acme, with its own oxygen steelmaking facilities based on iron made in a cupola, produced strip which was turned into strapping. Acme also had a materials-handling division. In 1968 Hoeganaes, a powder metallurgy company with a plant in Riverton, New Jersey, was acquired. In the late 1970s this operation was expanded through the addition of a plant in Gallatin, Tennessee.

In 1976 Interlake acquired Arwood Corporation, which was engaged in the manufacture of die and investment casting. In 1980 the Newport Steel Division, which was acquired in the 1950s, was shut down and in 1981 sold to the employees.

Interlake broadened its horizon in 1975 by acquiring Dexion, a European-based materials-handling company.

Current iron and steelmaking operations consist of two blast furnaces, one quite small with a capacity of some 350,000 tons annually, and the other considerably larger, with a capacity of over 600,000 tons annually. These are supported by two batteries of coke ovens. The iron from the blast furnaces is fed to two oxygen converters which have a capacity of approximately 900,000 tons per year. Steel is processed in a rolling mill, whose principal product is strip.

At the present, there are no plans for major changes, so that the company will continue its operation at a level consonant with business conditions. In early 1984 steel operations were at 75 percent of capacity.

Lone Star Steel

Lone Star Steel was incorporated in April 1942, at which time it consisted of a coke-oven battery and a blast furnace, both erected by the government in connection with the war effort. In the late 1940s, the plant was sold to private interests. It continued as an iron producer until 1953, when it became an integrated mill with the installation of an open-hearth shop and finishing facilities for pipe production. Initially the open-hearth shop consisted of four furnaces with a capacity of 550,000 tons annually. This was increased to 800,000 tons by the addition of another furnace in the late 1950s. In the mid-1970s, two 60-ton electric furnaces were added, which brought total capacity to approximately 1.0 million tons.

In 1966 Lone Star was acquired by Philadelphia Reading Corporation which, in turn, was acquired by Northwest Industries Corporation in 1968. Currently Lone Star is a division of Northwest Industries. Since steel production began Lone Star has been heavily and at times almost exclusively dependent on the pipe market, particularly the oil-country-tubular-goods market.

At the present time, facilities consist of a modern, well-functioning battery of coke ovens installed in 1979, one blast furnace with an annual capacity of approximately 800,000 tons, five open hearths, and two electric furnaces. The electric furnaces feed a continuous caster and were used recently when business did not warrant the operation of the blast furnace and open-hearth shop. They are also used in times of peak demand to supplement the open-hearth-shop output.

At present there are some options open to Lone Star to restructure its operations. One of these would be the installation of a basic-oxygen steelmaking shop to replace the open hearth, but this seems unlikely. The open-hearth shop is functioning well and has the advantage of using a high percentage of scrap when scrap is cheap. Further, the capital cost of replacing the open hearth would require a large investment which the company (after two poor years) is not in a position to make. The second possibility is the adoption of ladle metallurgy, which is far less costly and much more likely.

In the past few years, investments have been made to reduce costs and save energy. A ladle desulfurization process for pig iron has been introduced and a number of modifications have been made to the finishing facilities. In addition, a major research laboratory has been constructed.

In terms of product output, some 90 percent is welded pipe and 10 percent, which is used for drilling deep wells, is seamless.

The company does not plan any major capital expenditures in the next year or two. Thus, it will continue to operate with a capacity of 1.0 million tons that are dedicated overwhelmingly to the manufacture of pipe.

In terms of raw materials, Lone Star uses local ores which are sintered, as well as pellets which are purchased from Pea Ridge, Missouri. Coal is obtained from a mine located in Oklahoma.

Summary of Part I

The foregoing analysis of the integrated steel companies and their plants indicates the degree to which they have restructured their organizations and updated their facilities. Most of the plants have been materially improved with the installation of new equipment such as continuous casters designed to reduce costs, improve quality, and save energy. Obsolete equipment has been abandoned in wholesale fashion and in several instances relatively new facilities were shut down when they were unprofitable and showed little promise of improvement. The corporate structure has been changed by mergers and the acquisition and disposition of divisions and subsidiaries to fit long-range plans designed to increase efficiency and restore profitability.

In the move to increase efficiency, new facilities have been installed and plants have been combined, reduced in size, and in some instances closed down. In addition to the large investments for new equipment, a number of small, less significant expenditures have been made to improve operations. They include gauge controls and computer applications.

The restructuring and updating process, brought on in part by the steel depression, is now recognized as a continuing effort if the steel companies in the United States are to survive as competitive units.

Note

1. *8-K Form,* December 27, 1983, CF&I Steel Corporation, p. 3.

Part II
Restructuring the Steel Industry: Questions to be Considered

In the early 1970s (particularly in 1973 and 1974) the steel boom brought with it a tremendous sense of optimism to the integrated steel producers in the United States, since it was impossible for them to fill the steel demands of the economy which were strong not only in the United States but in many countries throughout the world. This led to projections that the industry in the United States would require 175 million tons of steelmaking capacity by 1980 and many of the companies planned to increase their capacities to participate in this growth. All of the integrated producers and many of the electric-furnace operators planned significant increases through the construction of additional plant and equipment. The total amount of added capacity projected was approximately 23 million tons involving twenty-nine companies. The facilities included both iron and steelmaking units as well as rolling and finishing mills.

This optimistic attitude was prevalent not only in the United States but worldwide as well. At the 1974 Annual Meeting of the International Iron and Steel Institute in Munich, most of the world's steel producers were represented and expressed unbounded confidence in the future as plans were discussed to add 240 million metric tons of steelmaking capacity in the world outside of the Communist orbit. This was in keeping with the opinions expressed at the Paris meeting of the same organization in 1970, when projections were made that the world would produce 1.0 billion net tons of raw steel by 1980. The optimism was short-lived as the downturn in the world economy and the collapse of the steel industry in 1975, followed by a modest revival in 1976 and another collapse in 1977, resulted in the cancellation of most of the steel expansion plans. In subsequent years, although there was a recovery in 1979 when the industry in the United States shipped over 100 million tons of product, there was not a revival of the euphoric spirit about the future of steel. Between 1977 and 1980, the integrated steel companies in the United States reduced their capacity by some 14.0 million tons, including two bankruptcies.

Since then, the steel producers have been at best cautious and at worst pessimistic about the industry's future. This has led not only to a moratorium on expansion but a further shrinkage of steelmaking and finishing facilities. Since 1981 over 21.0 million tons of raw-steel capacity were closed down by integrated companies in the United States. A number of steel producers, in addition to shrinking the size of their plants, have looked to other areas to supplement their steel income and protect themselves against cyclical fluctuations. However, most steelmakers, although staggered by the large losses, still believe that steel is essential, and that with a well-directed effort as well as an improvement in the marketplace, it can be profitable again. In keeping with this attitude, virtually all of the large producers have restructured their companies and eliminated as much of the high-cost, obsolete facilities as possible in order to compete in the future.

It is painfully evident to observers of the steel industry that this restructuring was necessary for survival and it is by no means concluded. In today's global competitive situation, the industry in the United States is operating too many plants and producing too many unprofitable products. Further, the flow of imports (often underpriced and subsidized) has made matters worse. Despite some restrictions on the European and Japanese imports of steel, about 20 percent of the U.S. economy's steel requirement came from foreign producers in 1983. To meet this situation, the U.S. steel companies must be competitive in their own market, and this requires reorganization of companies and their plants and a reasonable approach by all concerned to the solution of the trade problem.

14 The Steel Industry of the 1990s

The steel industry will change significantly in the remainder of this decade, from which it will emerge with a substantially different structure. There are a number of factors that have brought about changes thus far and will continue to force even more radical innovations in the future. These include:

1. The possibility of mergers, joint ventures, abandonment of obsolete, high-cost facilities, and diversification.
2. Competition from imports, which will continue to be a strong influence on the U.S. market.
3. The growth of the electric furnace in relation to the basic-oxygen converter and the continued development of minimills which are a considerable factor in the steel industry, affecting the integrated plants.
4. Declining employment as the integrated companies shrink in size and install more labor-saving facilities.
5. More attractive financing arrangements offered by equipment suppliers.
6. Increased use of purchased semifinished steel as a supplement to raw-steel production, rather than a replacement.
7. The formation of new steel companies from plants of significant size that are marked for abandonment or severance.

All of these factors will influence the future of the steel companies and will result in a smaller, more efficient industry. This shrinkage gives rise to another question: How much of the U.S. market will a smaller industry be able to supply?

An analysis of the above factors and influences should give some indication of what might be expected of the steel industry by the end of the 1980s.

15 Restructuring of the Steel Industry

Restructuring can take place in a number of ways: mergers, joint ventures, abandonment of facilities, diversification, and the adoption of new steel technology.

Mergers

The most effective type of restructuring is through mergers. Two or more companies, if merged, have one management; and decisions on plant rationalization can be made more readily than if a number of independent managements are involved, as in a joint venture. For example, the recent merger of J&L and Republic permits easier decisions for rationalizing plants in the Cleveland, Youngstown, and Chicago areas. Decisions can be made to eliminate parts of plants and combine facilities. This is particularly true in the Cleveland area, where both companies have plants producing sheets and where decisions will probably be made to eliminate parts of plants and combine facilities to form one unit. This unit will operate more efficiently and at lower costs than the two facilities. The same can be said of facilities at Youngstown and Chicago.

A merger that results in a unified, single management allows effective planning for the future by permitting the restructuring of plants and operations and the directing of capital expenditures in the most economical manner. It can lead to savings not possible if companies remain independent, for with an increase in size of plant and equipment, it is easier to prune away obsolete and high-cost facilities. In addition, selling and administrative functions can be combined, thereby reducing or eliminating costs.

Further, if the participants have product lines that complement each other, it eliminates the need for additional capital expenditures that could lead to overcapacity and a low rate of utilization. For example, if the proposed merger of Youngstown Sheet and Tube and Bethlehem in the late 1950s had been allowed, Bethlehem would not have found it necessary to build a grass-roots plant in the Chicago area.

A merger should, if it is properly constructed, result in a stronger company that is more able to compete and thus increase rather than reduce competition.

In judging the acceptability of a merger, the Justice Department places a great deal of emphasis on the concentration that will result in the marketing of a particular product. In respect to steel there has been a significant change in the factors establishing this competitive situation. In addition to the domestic companies that are competing for a particular market with a particular product, the competitive base has been substantially broadened. It now includes not only the domestic steel companies but also foreign companies supplying this market (whether or not there is a quota on imports) and other substitute materials. For example, in competition for the sheet business in the automobile market, the U.S. steel companies not only have each other but also more than a dozen active competitors from abroad. In addition, there are other materials, such as aluminum and plastics, that have cut significantly into what was once an exclusive steel market.

Another example is the container industry, which was once the domain of tinplate. In recent years aluminum has taken virtually all of the beer-can market and a significant segment of the soft-drink market. Other materials, such as plastics and impregnated paper, have claimed substantial shares. Consequently, the concentration from any merger should be judged not only in terms of the steel companies producing a particular product but should also include imported steel and inter-industry competition from other materials.

Joint Ventures

A second means of restructuring is through joint ventures. This is not as satisfactory as a merger and can best be applied to a new facility that two or more companies plan to bring into existence. An example is the current arrangement between Bethlehem and Inland to develop galvanizing capacity. It can be mutually decided where the facility is to be placed, who is to operate it, how much tonnage each will provide for the process, and how much product each will receive.

Joint ventures afford two significant advantages:

1. In a capital-intensive industry, such as steel, use of a joint venture involving two or more companies to install a high-cost facility reduces the required capital investment per company; and
2. Operating costs can be substantially reduced, since a jointly owned facility will usually function at a much higher rate of capacity than two or more such facilities that are independently owned. This reduces the depreciation charges and other costs per ton of product.

For example, a joint-ventured hot-strip mill operating at full capacity of 4.0 million tons would have a much lower depreciation cost per ton than two such mills operating at lower rates of capacity. Depreciation charges, reckoned on today's high-cost capital investment, could be substantial and, if cut in half, would make a big difference in operating costs.

Joint ventures, which have heretofore been limited to iron ore mining, should be extended to other basic steel industry operations, such as coke ovens, blast furnaces, steel-melting facilities and, possibly, some basic rolling operations. The product of the jointly owned facilities would be iron or semifinished steel and would be further rolled or formed at the wholly owned facilities of each company.

In years past, joint ventures have been advocated for the expansion of steel capacity, particularly where greenfield-site plants were involved for the production of semifinished steel. An example of this is the Tubarao plant in Brazil where three countries—Brazil, Japan, and Italy—are involved. However, at the present time, the need for modernization in the United States far surpasses the need for expansion, and with modernization costs at a high level, it would be well to consider joint ventures as a means of replacing worn-out and obsolete equipment.

Joint ventures on existing facilities are difficult to organize. When there are two or more managements involved, there is naturally a matter of self-interest on the part of each, since partners in the venture will want to retain as much as possible and give as little as possible to protect their own interests. By comparison, in a merger, there is not a multiplicity of interests but one unified interest.

Abandonment of Facilities

A third means of restructuring is the elimination of plants and the combination of facilities within an individual company. This has been carried out in most of the integrated steel companies. USS, as indicated elsewhere, has made drastic changes affecting its steelmaking facilities, reducing its capacity in the past ten years by more than 12.0 million tons. Bethlehem has closed down a major facility (with 3.0 million tons of capacity) at Lackawanna. Republic, in an action unrelated to the merger attempt, has abandoned a 1.0-million-ton plant in the Buffalo area. National has cut the capacity of its Great Lakes plant (near Detroit) in half and has spun off Weirton. Armco has shut down its Houston plant and is in the process of integrating as far as possible its Ashland, Kentucky, and Middletown, Ohio, operations. CF&I has abandoned its coke ovens, blast furnaces, and oxygen steelmaking shop.

As of early 1984, J&L has not made a decision for a radical reduction of capacity because it was awaiting the outcome of the merger with Republic. The merged unit will certainly proceed with some capacity reduction in the near future.

In addition to the reductions already made, there will be others in the next year or two that could result in a loss of 5.0 to 7.0 million tons of steelmaking capacity. These will eliminate high-cost, obsolete facilities and, to some extent, strengthen the companies involved.

These internal reductions are of some help insofar as they eliminate high-cost facilities, but they have not, in a number of instances, completely solved the problem. An indication of this is the fact that several companies are interested in further reducing their carbon steel business and, possibly, eliminating it through sale or closure.

Diversification

The fourth means of restructuring is diversification into nonsteel activities. This procedure has been adopted by several major companies including USS, Armco, National, and LTV. The remaining, major integrated companies that have not diversified are Bethlehem, Inland, and Republic. They have chosen not to follow this route but have concentrated instead on steel and closely related steel activities.

Diversification, if properly carried out, can mean security against the downside in the steel cycle. However, it must be pointed out that there are risks in diversification, and, in a number of instances, attempts have resulted in failure. Further, for a large company with billions of dollars in sales, diversification must involve a sizable operation or it cannot achieve its purpose. Opportunities are not presented every day, so this means of restructuring must await an attractive proposition. The sallies by the major companies into oil, financing, chemicals, and aluminum have proven quite successful, and although some companies have not done much in this area, they will have to evaluate any attractive possibility that presents itself.

New Technology

Restructuring requires the installation of new steel technology, which is discussed in chapter 21.

16 Competition from Imports

For fifteen years after the close of World War II, imports of steel were not a problem in the United States. Indeed, the United States was a net exporter of steel for all of the years after the war until 1959. In 1957, exports were 5.2 million tons as opposed to 1.1 million tons of imports. This changed in 1959, when the industry sustained a four-month strike, and steel users were desperate to obtain material. For the first time in the twentieth century, steel imports rose above exports on a three-to-one ratio, with imports at 4.5 million tons and exports at 1.5 million. This was considered a most abnormal situation because of the strike.

In 1960, imports maintained a slight edge over exports, as they reached 3.3 million tons as opposed to 3.0 million tons for exports. After that time, imports rose steadily, while exports remained within the 2.0- to 4.0-million-ton range, with the exception of two years (1970 and 1974) when they rose to 7.0 and 6.0 million tons, respectively. On the other hand, imports, due to lower prices and the triennial threat of a steelworkers' strike, increased to 18.0 million tons in 1968.

In 1969, both the Western Europeans and the Japanese imposed quotas on themselves and restrained their exports to the United States. This arrangement was to last for three years until the U.S. steel industry could invest enough in plant and equipment to bring it to a competitive position. The voluntary restraint agreement was renewed in 1972, but with the boom of 1973 and 1974, when steel was in great demand, the quotas were abandoned. Imports again became a problem in 1977 and 1978, when they reached an all-time high of 19.0 million and 21.0 million tons, respectively.

The basic problem in the late 1970s and early 1980s was the low price of imports due to the fact that they were subsidized by governments in a number of countries. In fact, government ownership between 1965 and 1981 had increased greatly in Europe and the Third World. For example, in 1968, there was no government ownership in the French steel industry, but by 1981, the industry was 70 percent government-owned. In Belgium, ownership rose from nothing in 1965 to 57 percent in 1981, while in the United Kingdom, it rose from 8 to 76 percent over the same time period. In the United Kingdom, this was not a gradual increase but an abrupt change which took place in 1967 with the nationalization of most of the British steel industry. Since 1970 Third World nations have built steel industries that in most cases are government-owned.

Imports take a very significant share of the American market and recently have been at 20 percent or better. Further, the source of imports has been changing, as the steel industry in the Third World grows up and has millions of tons of steel to export. This is particularly true of Brazil and South Korea where there has been a dramatic increase in steel capacity in the last ten years. South Korean output rose from slightly less than 500,000 metric tons in 1971 to almost 12.0 million tons in 1983, while Brazilian production rose from 6.0 million metric tons in 1971 to a high of over 15.0 million tons in 1980 and, although it declined somewhat in the following two years, recoverd to 14.6 million in 1983. Exports of both of these countries to the United States have increased appreciably in the last few years. For the five-year period from 1978 to 1982, South Korean exports to the United States averaged about 1.0 million net tons. In 1983, the figure rose to 1.7 million. During the 1979 to 1982 period, the average imports from Brazil were somewhat over 500,000 tons. In 1983, they increased sharply to almost 1.3 million tons.

In late 1982, an agreement was reached with the European community to limit exports to the United States to approximately 5 percent of the market. Shortly thereafter, the Japanese agreed to restrain their exports. Thus, in 1983, exports from the Common Market amounted to 4.1 million tons, down from 5.6 million in 1982, while the Japanese exports were 4.2 million, down from 5.2 million in 1982.

It is a basic fact of economic life for the American steel industry that imports will continue and will take at least 15 percent of the market for the years ahead. Currently, there are two efforts to limit global imports to that figure. One is a bill introduced in the House of Representatives of the U.S. Congress, entitled the Fair Trade in Steel Act of 1984. It states that:

> The Congress acknowledges that the steel industry in the United States is critical to the national defense and the maintenance of a strong industrial economy which employs millions of workers and sustains the Nation's prosperity. The Congress finds that international trade in steel over the past two decades has been plagued by unfair trade practices such as dumping, subsidization, industrial targeting, discriminatory bi-lateral agreements, selling below cost, and other predatory practices . . . The result has been a substantial injury to the domestic steel industry . . .[1]

In order to take remedial action that will prevent further decline in the domestic steel industry, the bill states:

> It is, therefore, declared to be a policy of Congress that access to the U.S. market for foreign-produced carbon, alloy, and specialty steel-mill products, should be on an equitable basis to safeguard the national security, insure orderly trade in steel mill products, reduce unfair trade in steel mill products, and alleviate U.S. balance-of-payments problems. In order to accomplish

> this policy, it is deemed necessary and appropriate to limit imports of steel mill products to approximately 15 percentum of apparent domestic supply for at least a five-year period.[2]

The bill covers virtually all steel products, including semifinished.

The second is a petition filed under Section 201 of the Trade Act of 1974 by the United Steelworkers of America and Bethlehem Steel Corporation charging serious injury from imports. In reference to imports and import pricing, the petition states:

> Such massive price depression and suppression caused by the availability of imported merchandise constitutes by itself an important cause of the serious injury suffered by the domestic steel industry . . .
>
> No other cause of injury is more important than increased imports and the resulting increased pricing pressure from imports . . .
>
> By this petition, the United Steelworkers of America and Bethlehem Steel Corporation seek the limitation on imports into the United States of carbon and alloy steel products (including stainless steel and tool steel), product by product and country by country, to a total of less than 15% of apparent domestic consumption for all products for at least the next five years.[3]

As in the case of the bill to be presented to Congress, the petition also includes semifinished steel among the products to be limited.

How successful either or both of these attempts will be remains to be seen and will not be known until at least the fourth quarter of 1984. It is interesting to note that both approaches are willing to concede 15 percent to imports, thus, theoretically, leaving 85 percent of the U.S. market to be satisfied by domestic producers.

In respect to steel quotas on imports, it should be pointed out that the European Economic Community (EEC) has established them for countries shipping steel into the EEC from outside. These quotas apply to Western European countries outside of the EEC, such as Spain, and all others around the world that ship steel into the EEC. Japan has had a quota for quite some time although it has not filled it.

There are no set quotas against imports into Japan, but recently a strong protest was made about imports from South Korea and Taiwan, which have amounted to 3 percent or 4 percent of total Japanese consumption.

In some Third World countries, the steel industry controls imports either directly or indirectly, directly by ordering and selling them and indirectly by influencing government decisions as to what products may be imported and how much. This is practiced in a number of Latin American countries.

The quota that has been negotiated by the United States with the EEC is firm in terms of tonnage exported to the United States; however, the self-

Restraint which the Japanese adopted and practiced in 1983 seems to be changing when one considers the imports from Japan in December of 1983 and January of 1984. In the latter month, imports from Japan were 530,000 net tons, which if annualized would equal 6,360,000 for the year. In February of 1984, imports from Japan amounted to 622,000 tons.

The Japanese and Western Europeans have long been the principal steel suppliers to the American market and have made every effort to continue their participation despite some limitations. Both groups have sales organizations functioning in the United States and have had longstanding relationships with customers, so that they refer to "their share" of the U.S. market. The total share for all countries will be at least 15 percent of the U.S. market and, if the two attempts to limit imports fail, the figure may be significantly higher.

The acceptance by the EEC of a quota was conditioned on U.S. steel companies dropping their antidumping and countervailing duty suits against a number of companies in Western Europe. There were no significant suits against the Japanese; however, they adopted a policy of restraint in order to reduce the possibility of legislative quotas by the U.S. Congress.

The quotas for the EEC and the self restraint of the Japanese were expected to reduce imports. However, this did not happen, for imports in 1983, the first full year of this restraint, were not only not reduced but actually were slightly higher than those of 1982. This was due to a significant increase from those sources not under quota or restraint, such as the Third World countries and Canada. Total imports in 1983 were 17.007 million tons as opposed to 16.663 million tons for 1982. Common Market shipments to the United States were down by 26.5 percent and the Japanese down by 18.3 percent. However, Canada increased by 29 percent and the "All Other" group, principally Third World countries, increased by 57.1 percent. As a consequence, the Western Europeans and Japanese have expressed resentment over the fact that the reductions they have made in tonnage have been taken up by other countries. This may well explain the increase in Japanese imports in December 1983 and January 1984. To remedy this situation some type of global arrangement is needed.

This bill, submitted to provide worldwide quotas as well as the petition filed under Section 201 of the Trade Act of 1974, caused considerable concern among all of the countries exporting steel to the United States. The Third World countries feel that it will affect them significantly, since there is a general impression that 15 percent will be allocated, 5 percent to the European Common Market (which is approximately what it has under the present quota), 5 percent to the Japanese (who will want to maintain parity with the Common Market), and 5 percent to the rest of the world. This formula might have been acceptable in 1968 and 1970; however, since that time, the world outside of the European Community and Japan has greatly increased its steel capacity. The figures for Brazil and South Korea are given above.

The Third World countries feel that a 5 percent participation, which includes the Canadians, would limit them excessively in their exports to the United States and force a reduction that they are reluctant to accept. Some maintain that a more equitable distribution would be 4 percent for the European Community and 4 percent for Japan, since those industries are shrinking and 7 percent for the rest of the world, where the steel industry is growing. This would not be acceptable to the Europeans or the Japanese.

If the bill passes or the Section 201 Petition is accepted, some formula will probably be worked out to accommodate Canada and the expanded industries of the Third World, perhaps by increasing the quota to 17 percent. The South Koreans have been particularly vocal in this respect, maintaining that they have an expanded industry with modern equipment and low-cost production, and their exports to the United States at favorable prices will benefit the U.S. consumers.

The effect of imports on the steel companies in the United States will be a permanent loss of at least 15 percent of the total market. This is an acceptable situation; however, what is difficult and unacceptable is a 20 percent or higher penetration of the U.S. market by imports at prices that, in a number of instances, are below the cost of production.

In January 1984, imports absorbed 25 percent of the U.S. market which resulted in strong protests by the steel industry to the government at the highest levels. The U.S. companies maintained that they cannot compete with subsidized steel, which constitutes a large segment of imports. In a letter to President Reagan, the Chairman of USS Corporation and of the American Iron and Steel Institute, David M. Roderick, said: "In all candor, Mr. President, I must tell you we are at the peril point."

The impact of foreign competition has been substantial in respect to many of the companies and devastating for some. There is considerable concern that a portion of our remaining steel capacity may be closed in the face of low-price imports. An example of this was the Houston mill of Armco, which had to close down in late 1983. There is also concern that the loss of more capacity, if it drops below the total of 120 million tons, will generate the need for imports as a supplement to U.S. steel supply rather than in competition with it. If this occurs, there is a distinct possibility that prices of imports will increase appreciably, and the consumer who has benefited from low-cost imports will have to pay higher prices for a long period. This is precisely what happened in 1973 and 1974, when there was a strong demand in the United States beyond the ability of the domestic industry to supply it. Plate imports that sold for $119 per metric ton FOB a Western European mill in January 1972 rose in price by July 1974 to $440 per metric ton.

If sufficient capacity is closed down in the United States in the near future (and this could be in two to three years from 1984), imports will be needed. The prices will be higher than the domestic level, not only on a temporary one- or two-year basis but for a much longer period.

Imports have contributed to the restructuring of the U.S. steel industry insofar as they have been instrumental in forcing the shutdown of facilities, some of which were high-cost and obsolete. Imports will continue to keep pressure on the U.S. steel producers even under a restricted 15 percent quota. This size quota would represent 15.0 million tons or more, and such an amount will continue to offer price competition to the U.S. producers. A number of products imported from Europe under the quota have been priced below U.S. domestic prices and have fostered competition.

The steel producers outside of the United States in other than a boom period will compete vigorously for the U.S. market, the largest and least protected in the world. This competition must be rational and not destructive, so that the domestic producers and importers can operate at a profit. Because of the unstable condition that has prevailed in this market, it appears that some regulations are necessary to restore order. Individual suits on dumping and countervailing duty have their place; however, they can only rectify specific cases and require a great deal of time and expense for both parties involved. The present proposals, perhaps in a modified form generally acceptable, can provide a solution to a critical problem.

Notes

1. Fair Trade in Steel Act of 1984 (H.R. 4352), p. 2.
2. Ibid., p. 6.
3. Petition by United Steelworkers Union of America and Bethlehem Steel Corporation filed under Section 201 of the Trade Act of 1974, pages VIII and IX.

17 Growth of the Electric Furnace and the Minimill

In the past two decades much attention has been given to the growth of the basic-oxygen converter as a steelmaking process in the United States. In a matter of twenty years it displaced the open hearth, which had been firmly entrenched as the principal steel-producing process. In the 1950s the open hearth produced 90 percent of the steel in the United States, with the Bessemer and electric furnace contributing the other 10 percent. During the 1960s and 1970s the basic-oxygen converter was installed by almost all integrated steel companies and, with few exceptions, in most integrated steel plants. This was heralded as a great revolution in steelmaking which, indeed, it was. However, not much attention was paid to the electric furnace, which also had a significant rise in the last thirty years. In 1950, it accounted for 6.0 million tons (or approximately 6 percent) of steel output; by 1969, electric furnace output had increased to 20.2 million tons (or 15 percent) of U.S. steel production. In the following decade this process experienced a rapid growth in production to 34.0 million tons, or 25 percent of the nation's steel output by 1979; in 1982 it accounted for 31 percent.

The electric furnace during World War II and for a period thereafter was considered a specialty steel producer, primarily for alloy and stainless steel. Since 1960 there has been a sharp change in the output of this furnace, so that now 70 percent of its production is in carbon steel.

The electric-furnace rise as a producer of carbon steel can be attributed to a number of factors including such innovations as: the introduction of ultra-high power, the injection of oxygen, and the application of water-cooled side panels and roofs. These improvements have increased the capacity of many units, reduced the amount of time needed to produce steel, and prolonged the furnace lining life. In addition, the relatively low capital cost required for the installation of these furnaces led to the development of minimills as well as their adoption by a number of integrated producers.

The integrated steel companies have over eighty electric furnaces located at twenty-eight plants throughout the country, with a combined capacity of 20.5 million annual tons. Some of these furnaces are quite large, such as the two 350-ton units at J&L's Pittsburgh Works with an annual capacity of 1.8 million tons; USS Corporation has four electric furnaces, each with 220-ton heat capacity at its plant in Baytown, Texas; Bethlehem has electric furnaces at its Steelton plant and has recently installed two 185-ton furnaces at its Johnstown plant.

The integrated companies have installed electric furnaces partly because the capital cost is much less than that required for a blast-furnace/basic-oxygen combination, which often includes additional coke-oven capacity, and also because electrics can add smaller increments of steelmaking capacity. Recent estimates indicate that comparable electric furnace capacity can be installed at one-third the cost of the coke oven-blast furnace-BOF combination. Also, electric furnaces reduce the pollution problem since they eliminate the coke oven which is a necessary adjunct to BOF production and the prime cause of pollution.

In the last ten years, electric furnaces have been installed whereas a number of basic-oxygen converters have been withdrawn. Currently, the electric furnace capacity in the United States is approximately 48.0 million tons. Further, the electric furnace can be expanded without installing additional units; this is accomplished by installing water-cooled side panels which have increased the capacity of furnaces by 10 to 30 percent.

In between the minimills and the integrated producers, there are a number of companies with large electric furnaces that produce heavy steel products. Examples of these are Northwestern Steel and Wire and Lukens Steel. Northwestern produces steel entirely by the electric furnace process and has a capacity of 2.5 million tons per year from the three largest electric furnaces in the world; each is capable of producing 400 tons per heat. The finished products include wide flange beams. Lukens, likewise, depends on electric furnaces for its steel and rolls, among other products, heavy plate on one of the widest mills in the United States.

Problems

There are some problems connected with electric furnace steel production. Because scrap (which is used almost exclusively as a feed material) contains a number of contaminants (many of which remain in the final product) there are limitations on the type of steel that can be produced. This is particularly true of steel sheets for deep-drawing applications. Much has been done to solve this problem; however, it still poses a difficulty. Thus, the application of the electric furnace process for quality carbon steel will continue to be limited by the quality of the scrap used in the charge. In some plants the problem has been greatly reduced by using larger amounts of internal revert scrap—much of which is quite free of contaminants, as well as by the careful selection of purchased scrap.

Another difficulty facing the electric furnace operator is the increase in the cost of power. The average consumption of electrical power per ton of steel produced is 490 to 500 kilowatt-hours. Ten years ago electric power could be obtained for less than one cent per kilowatt-hour in many locations.

In the past few years, however, the price has tripled or quadrupled in many locations and in some, particularly in the Chicago area, it has risen to six to seven cents per kilowatt-hour.

Future Growth

The electric furnace has grown very rapidly, but this pace is not expected to continue. However, this process will grow as a percentage of steelmaking capacity in the years ahead, while the basic-oxygen converter, in terms of tonnage potential, may decline or at best remain stationary. By 1990 the electric furnace, if sufficient acceptable scrap is available, could produce as much as 44 to 46 percent of the steel made in the United States. It will challenge the BOF process but probably not surpass it.

The integrated producers will not install any more basic-oxygen capacity, while some will increase their ability to produce steel by the electric furnace process. An indication of this trend is the recent actions taken by CF&I at Pueblo and USS Corporation at South Works: In both instances the basic-oxygen process was eliminated and the plant left to rely on electric furnaces for steel output. This, however, is not expected to occur on a large scale in the near future.

The electric furnace has helped and will continue to help restructure the integrated steel companies in the late 1980s and 1990s, particularly if the scrap problem is solved. The basic-oxygen converter will account for most of the steel produced for deep-drawing applications since it uses pig iron, which is relatively free from contaminants.

Minimills

Minimills are small steel producers, many of which came into existence in the 1960s and early to mid-1970s. They were characterized by steelmaking capacities ranging from 50,000 tons per year to 150,000 tons and usually consisted of one or two small electric furnaces dependent on scrap. Very often they employed continuous casting, while the finishing facilities consisted of a bar mill which produced, in the early stages, concrete reinforcing bar. Later, small angles, channels, flats, and squares as well as smooth bars were added. At the outset, minimills were characterized by limited size, product range, and geographic market.

In 1970 there were over forty of these mills in operation throughout the country. The investment required for their construction in the 1960s was relatively low, ranging from $5.0 million when second-hand equipment was installed to $8.0 million to $10.0 million for new equipment. In recent years, the latest mills installed had a much larger capacity, ranging up to 500,000 tons and higher, requiring an initial investment of $60 million to $70 million.

Many of the small mills installed in the 1960s were expanded in the 1970s by the construction of an additional furnace or the replacement of small furnaces with larger ones. Very often a 30- to 40-ton furnace would be replaced by a 70-ton to 80-ton unit. About the same time, minimills reduced their output of rebar and devoted most of their production to small structural angles, channels, smooth bars, and narrow flats. More recently a relatively small number of mills, which, however, represent significant tonnage, have gone into the production of special-quality bars.

In 1984 the term *minimill* has lost its original meaning, for many of these plants due to expansion have more than 500,000 tons of capacity while some have reached the 750,000- to 1.0-million ton levels. For example, Chaparral in Texas is often referred to as a minimill and did begin as one; however, it has expanded and now has a 1.0-million ton capacity and is larger than Laclede Steel, which was never referred to as a minimill. In recent years an attempt has been made to change the name from minimill to market mill, since most of these mills serve a limited geographic market.

The market mill or minimill has not been defined but rather described by Edward L. Flom, Chairman of the Board of Florida Steel, a company that operates five small steel plants. In a speech before the American Iron and Steel Institute in May 1983, he said:

> . . . A market or mini-mill is a *concept* . . . the concept of a steel-producing facility with relatively low capital costs. It is modern and efficient equipment that produces sufficient hot tons to be rolled into a tailor-made finished product and then sold in a closely defined market with maximum flexibility in pricing, production, equipment advances, and particularly total employee participation. Each criterion is important but overall FLEXIBILITY is the key.

Because of flexibility and low capital costs, as well as concentration on a few specific products, these mills have been able to provide certain products at a lower cost than many of the integrated mills. As a consequence, during the past decade, a number of integrated mills have dropped out of the production of minimill products.

Without question, the minimills, which now number over fifty throughout the country with a capacity of more than 20.0 million tons, have established themselves as permanent participants in the steel industry. The minimills for a number of years were limited in their product lines; however, this has changed significantly in the last decade as they have moved into the production of special quality products. Examples of these include the North Star mill located at Monroe, Michigan, near Detroit, which has a capacity of some 400,000 tons and makes a variety of alloy bars. Quanex Corporation operates a mill at Jackson, Michigan, with less than 200,000 tons of capacity, that produces special-high-quality steel bars used in the forging industry and for the production of seamless pipe and tubes. The plant is highly efficient and one of the most modern of its type, not only in the United States but in the worldwide steel industry. At present, Quanex has a second such mill under construction at Fort Smith, Arkansas.

In the past year, there has been considerable discussion concerning the possibility of minimills producing relatively narrow steel sheets. It will take a few years before such a development can materialize, since it will require the casting of thin slabs, approximately one-and-one-half inches thick, and the installation of a rolling mill with either one or two stands to roll these slabs to sheets. The product, if produced, will most probably be confined to hot-rolled, narrow sheets and strip.

Several minimills have in recent years moved into the production of wire rods and now dominate that product. These include Raritan River Steel Company in New Jersey, Georgetown Steel Corporation in South Carolina, and the Texas plant of North Star Steel, formerly Georgetown Texas. These mills can produce wire rods with less than two man hours per ton and, consequently, have much higher productivity than the integrated mills, which have higher man-hour-per-ton requirements. As a consequence, the smaller electric-furnace mills can and have sold certain types of rods at prices that could not be matched by the integrated mills without considerable loss. Under these circumstances, several of the integrated steel companies, including USS, have dropped out of the rod business. In addition to rods, minimills have forced many of the integrated companies to restructure to the point of dropping out of other product lines, such as bar-size angles and channels. Minimills have also gone into the production of seamless pipe.

This penetration into what was to a great extent integrated mill territory will probably increase in the years ahead. However, there are a number of products that minimills will not be able to produce, such as wide sheets, and particularly cold-rolled sheets, as well as heavy wide plates.

In general, these small mills have functioned profitably; however, there were a number that have not been successful. Profitability has varied considerably and will continue to do so. A number of minimills have gone into bankruptcy in the past five years, while others have prospered. Three successful companies—Florida Steel, Nucor, and North Star—have several locations. Florida Steel has five plants, while Nucor and North Star each have four plants. Aggregate capacity for each of these companies is about 2.0 million tons, with one single mill having a capacity in excess of 500,000 tons.

During the past decade several of the integrated plants have considered the possibility of constructing a minimill or acquiring an existing one. Armco bought Pollack Steel, a minimill in the Cincinnati area, and sold it as well as another small electric furnace plant it had built at Sand Springs, Oklahoma. None of the major integrated companies operates a separately located minimill, although the electric furnace at Inland with its own caster and bar mill has been referred to as Inland's minimill within an integrated mill. This is indeed a misnomer since these facilities, although they form a unit, rely on the integrated mill for many of their services.

Since their inception in the early 1960s minimills have been considered a thing apart from the steel industry—a segment not quite integrated with the industry but something unusual. During the last ten years there has been a gradual but significant change in this concept. The minimills are now an integral part of the industry, and some of them have grown in size so that their tonnage is greater than a number of integrated plants. This was particularly true in 1982 when steel production fell dramatically by 40 percent. At least three of the minimills surpassed a number of the integrated mills in production. These were Nucor, which ranked tenth; Korf, which ranked eleventh; and Florida Steel, which ranked fourteenth.

The lines of definition of a minimill versus other steel plants have become, in many respects, so blurred as to be meaningless. Some integrated plants have now shrunk into what is considered a minimill. For example, CF&I at Pueblo, which has closed down its blast furnace and oxygen steelmaking capacity will operate in the future electric furnaces with a capacity of about 650,000 to 700,000 tons; this is comparable to many plants that have been designated minimills.

However, if one maintains that minimills should be defined by product, restricting them to the rebar and small structural items, the CF&I mill producing rails and seamless tube would not fit this category. In terms of size and technology; however, CF&I is definitely in the minimill category. Another example is the USS Corporation mill at South Works in Chicago. It has recently abandoned its blast furnaces and oxygen converters and, in the future, will operate electric furnaces only. The mill has a capacity of somewhat less than 1.0 million tons, which in terms of size would put it in the minimill category. However, its product line, wide flange structural beams, definitely is not a minimill product. Still other examples include a seamless pipe mill constructed at Youngstown, Ohio, with about 300,000 tons of capacity annually (again, not a traditional minimill product). The Johnstown plant of Bethlehem, which was formerly an integrated plant, is now a 1.2-million-ton electric furnace operation producing bars as well as semifinished billets for other Bethlehem plants.

As far as the future is concerned the so-called minimills or market mills—which have grown from a very small portion of the U.S. steel industry in 1960 to a significant 22 percent by 1983 (depending on what plants are included)—will continue to remain strong. They have a number of advantages, not the least of which is the personal employer–employee relationship. With a small work force, top management frequently knows virtually all of the employees; thus an atmosphere of mutual trust and confidence can be developed, which is a great asset in solving personnel problems.

The term *market mill* is losing its usefulness, for these mills are no longer confined to a small geographic area in relation to market, but in a number of instances sell on a national basis. Chaparral, for example, markets its products in forty states.

In terms of future growth, although the rate will decelerate considerably from the experience of the 1960s and 1970s, it will continue not so much from the construction of new mills but rather through expansion of existing mills. As previously indicated, larger furnaces are being installed: one company replaced a 45-ton furnace with a 90-ton furnace; another replaced a 30-ton furnace with a 65-ton furnace. In addition, the adoption of new technology, such as water-cooled side panels, increased electric power, and injection of oxygen, will give these mills more tonnage.

The small electric furnace, together with the continuous caster and a specialized product with limited tonnage will challenge the integrated mills with blast furnaces, coke ovens, and basic-oxygen converters so that by 1990 the former will represent 30 percent of the industry's output. This is particularly so when one considers the change that has taken place with the abandonment of integrated facilities in favor of the electric furnace.

If those plants that have been converted from integrated to electric furnace are added to the smaller electrics the total output is even more imposing. If long-time electric furnace operators such as Lukens, Northwestern Steel and Wire, Laclede, and Phoenix, are also added, output could reach 45 to 48 percent of the U.S. total by the early 1990s. This is quite possible when one considers that additional basic-oxygen capacity may be closed down. In fact, in the restructuring of the industry the facilities to be eliminated are rarely electric furnace operations; instead, they are open hearth and basic-oxygen furnaces.

The electric furnace has replaced and will continue to replace some BOF installations; however, in terms of total tonnage it will not replace the BOF, but will reduce its supremacy as the leading steelmaking process.

18 Purchase of Semifinished Steel

In the spring of 1983 USS Corporation and British Steel Corporation (BSC) began discussions about the possibility of supplying slabs to the Fairless Works from BSC's Ravenscraig plant in Scotland. This would have resulted in the shutdown of the iron and steelmaking facilities at Fairless and thus provoked a vigorous reaction from the United Steelworkers. Not only would the proposed arrangement have resulted in the loss of 1800 to 2000 jobs; the union also maintained that other companies might follow this example and import large tonnages of semifinished steel. This would result in a reduction in U.S. raw steelmaking capacity, a loss of jobs, and a greater dependence on foreign steel.

USS countered with the argument that if the company could not procure slabs from outside, thereby allowing it to shut down obsolete melting capacity at the Fairless Works, the plant would be closed down entirely within a few years, with the loss of many more jobs, since USS did not have the funds to modernize the facilities. Thus the arrangement to procure slabs made by modern facilities abroad would permit the plant in its finishing sector to become competitive, and some 4800 jobs in the rolling and finishing-mill areas of the plant would be saved.

The discussions between USS and BSC came to nought because BSC was not able to supply slabs at an economical price. However, the incident focused attention on the balance that exists or does not exist in a number of steel companies between melting and finishing facilities. The question then was asked: If there is an imbalance in favor of finishing facilities, would semifinished steel be procured from foreign sources to remedy the situation?

With few exceptions in the past decade, most of the plants, particularly the integrated ones, had a reasonable balance between melting and finishing facilities. However, with cutbacks in production, more melting capacity has been taken out of operation than finishing capacity. For example, National Steel at its Great Lakes plant eliminated a basic-oxygen shop with 3.0 million tons of raw steel capacity, while its 80-inch hot-strip mill remained in operation with a potential to roll more steel than the remaining melting units at the plant could produce. The same is true of Lukens Steel, which purchased the Alan Wood plate mill but not the Alan Wood steelmaking facilities. Thus Lukens has a mill with approximately 800,000 tons of rolling capacity and no steel to put through it. CF&I has just eliminated 1.3 million tons of steelmaking capacity but has maintained its pipe, structural, and rail mills to be fed by electric furnaces with less than 700,000 tons of capacity. Kaiser Steel's plant, which has been shut down, can only be reactivated if 1.0 million tons of semifinished steel can be obtained, since it has been reduced to a finishing facility with no production of iron and steel. These are but some examples of a permanent imbalance that has developed in the past few years.

There are also other plants that have more finishing than melting capacity, such as: Sharon, Armco at Middletown, and occasionally, Rouge Steel. This imbalance is particularly true of many plants with new hot-strip mills that are operating below their potential. In addition, a number of blast furnaces, steel furnaces, and finishing facilities have been shut down due to depressed business conditions. If the economy improves and orders increase, it is much easier to reactivate a rolling facility than a blast furnace and steelmaking facilities. With an increase in demand, particularly if it is not considered long-term, it makes more economic sense to obtain semifinished steel and operate the rolling mills than to start up blast furnaces and steelmaking facilities to produce iron and steel which will, in turn, support the production of semifinished steel. Consequently, under these conditions, the demand for semifinished steel in the United States will continue to be strong and may even increase. On the other side of the equation, there are very few plants with an excess of steel production over and above their finishing capacities. The most notable is Inland Steel.

An indication of efforts to remedy the problem by bringing in foreign semifinished steel can be had by examining the import figures for the 1981–1983 period. During these years semifinished imports increased substantially; they averaged almost 800,000 tons per year for the period as opposed to less than 300,000 tons for the eight years prior to 1981. During 1984 and 1985, more slabs, billets, and blooms will be sought from abroad; however, there are very few mills throughout the world that, under normal circumstances, have a steady excess of semifinished steel for sale. One of these is the recently inaugurated plant at Tubarao in Brazil, which can produce 3.0 million tons of slabs that are for sale, since the mill has no finishing equipment. The plant is a joint venture involving Siderbras, the Brazilian Steel Company, Kawasaki Steel Corporation of Japan, and Italsider of Italy. The plan to build it was conceived in the early 1970s when steel was in short supply and projections for future growth were highly optimistic. Originally the Japanese and Italians planned to take very large tonnages of semifinished and roll them on finishing facilities in their respective countries. By 1983, when the plant was finished, the entire steel picture in the world had changed, and there was a definite oversupply. As a result, the Italians and Japanese have no need for the slabs at home and are anxious to sell them on the international market.

Another steel company with an excess of semifinished is the Svenskt Stal Aktiebolog (SSAB). This was formed recently from a merger of the three largest carbon steel producers in Sweden and has more than 700,000 tons of semifinished steel, which it has offered and will continue to offer for sale.

Presently there are integrated mills throughout the world that are willing to sell slabs on a limited short-term basis, but do not wish to enter into long-term, fixed-price contracts, since the sale of slabs is less profitable than that of processed steel. Several of these companies are currently selling semifinished steel in the United States in substantial tonnages; however, the reluctance to enter into long-term commitments indicates that if there is even a slight improvement in the worldwide steel picture these suppliers would either reduce their sales of semifinished steel or completely eliminate such sales. The reluctance to sell semifinished steel is particularly true of Japan which, in 1982, exported a little over 7000 tons of semifinished steel to the United States out of a total of 716,000 tons. In 1983, out of a total of 823,000 tons of semifinished steel imported by the United States, the Japanese accounted for slightly over 1000 tons.

The countries providing semifinished steel in significant quantities in the last two years include Canada, West Germany, South Africa, France, Brazil, The Netherlands, Sweden, and Spain. Canada has contributed very heavily, accounting for more than one-half of the total in 1983. In Europe there are several companies willing to sell semifinished steel on a limited-time basis, since they have more melting capacity than they are allowed to roll under the EEC quotas. These companies are in France, West Germany, the United Kingdom, and Holland. The willingness to sell slabs is further nourished by the fact that the Western Europeans can get a higher price for slabs in the United States than they can get for plates in some other countries, including China.

With the reluctance on the part of most current semifinished suppliers to enter into long-term contracts, it would be irrational for a company in the United States to abandon its iron and steelmaking facilities in the hope of procuring semifinished steel on a permanent basis. The only possibility for making a permanent arrangement to secure semifinished steel would be through an investment in the supplying company.

Semifinished steel will be used in increasing quantities in the United States as a supplemental material rather than as a replacement. Virtually all the plants purchasing semifinished steel make most of their own steel, so that semifinished steel is needed only in times of a sudden increase in demand or at peak operation. However, plants such as Kaiser Steel and the newly created rolling mill at Ohio River Steel Company must have semifinished in order to operate. These companies, as well as a number of others, are concerned about the attempts to limit semifinished imports. They maintain that only steel companies import semifinished steel, and thus it is not a competitive product. However, the other opinion is that if the semifinished steel comes in at low subsidized prices, it gives the importer an advantage.

The demand for semifinished steel in the United States will persist; however, there is a problem concerning its supply.

19 Declining Employment in the Steel Industry

At the peak of the steel boom in the United States in 1973—when 151 million tons of raw steel were produced and 111 million tons of finished product shipped—the average number of steel industry employees was 509,000. In the spring of 1981, with the industry operating at 90 percent of capacity, the average number of employees had dropped to 391,000. Thus, there was a reduction at the same operating rate of 118,000 jobs, or more than 20 percent of the total. In 1982 to 1983, with a sharp drop in activity, the number of employees fell off considerably; the average for the year 1982 was 280,000, which declined to 243,000 in 1983. While the 1981 figure of 391,000 jobs represented a permanent loss, the reduction to 243,000 in 1983 included a large number of those on temporary layoff. When the industry returns to an operating level of 85 percent or more, a number of these employees will be recalled. Thus the permanent force at that level should be somewhere in the neighborhood of 310,000 to 325,000. This is a drop of about 40 percent from the 1973 level and represents jobs lost permanently as a result of the abandonment of a significant segment of the steel industry's productive capacity.

Between 1981 and 1984, over 21.0 million tons of raw steelmaking capacity were abandoned by the integrated steel producers. There will be further reductions by 1985, resulting in corresponding reductions in employees. Virtually every company has been affected, since all but Inland and J&L have reduced capacity significantly. The bulk of the cuts have been made, but some others can be anticipated.

This drastic cut in employment has had a devastating impact on certain areas, while others have been able to absorb it more readily. The areas known as steel towns have been severely affected. Communities such as Youngstown, Ohio; Johnstown and Aliquippa, Pennsylvania; Lackawanna, New York; and Pueblo, Colorado, have been hard hit. For example, the J&L plant at Aliquippa in 1980 and 1981 had approximately 9500 employees; by 1984 this figure had been reduced to 3500. Other areas—particularly the larger cities such as South Chicago, Detroit, and Cleveland—have felt the impact, but due to the size of the communities and the variety of industry it has not been as devastating as it has for the smaller communities.

The closure of plants by USS Corporation was announced in December 1983; this will result in the permanent loss of 15,000 jobs, although 10,000 employees were on layoff at the time of the announcement. Armco reduced total employment by 15,000 during 1982, while the closure of the Bethlehem Steel plant at Lackawanna eliminated some 5000 jobs. The facilities involved have been abandoned, so the loss is permanent, as was the case with other steel companies abandoning plants.

Another negative influence on employment in the steel industry is the installation of new technology that requires fewer employees. For example, a continuous caster eliminates a number of steps in steel production and, consequently, some employees.

There is virtually no hope that the industry will expand its operation in the next few years. Thus there will be no opportunities for additional jobs. In one instance—the USS plant at Fairfield, Alabama—the facility has been reactivated after being closed for over a year and a half; thus, a number of jobs that had been lost were restored. There is little hope that this will happen in any other area. Therefore the employment situation can be described as grim.

In order to prevent the closure of a number of facilities, wage concessions that reduced employment costs by about 9 percent were made by the union in March of 1983. Additional concessions were sought by a number of companies, but in several cases the union refused to make them. The result was the closure of facilities. In several plants the concessions made were significant enough to allow continuation of the operation. For example, at McLouth a $5-per-hour concession was made by the union; at Rouge it was $4.50 per hour; a $5-per-hour concession was also made at CF&I in Pueblo in 1982, and an additional $1-per-hour concession was made there in 1983. However, at CF&I this did not prevent the permanent closure of the plant's coke ovens, blast furnaces, and steelmaking capacity. It should be pointed out, however, that these facilities had already been closed down on a temporary basis when these concessions were made.

Further concessions on an industry-wide basis are not likely; however, in the case of individual companies or plants, it is more than possible that the employees will be willing to take a reduction in wages or work-rule changes when the permanent loss of their jobs is the alternative. The outstanding example of this is the Weirton Steel Company, where the employees bought the plant and took reductions of 32 percent in their wage rates to reduce costs, so that the plant would be economically viable. This cut was taken because the employees bought the plant. They would not have done this if National had continued to own and operate the facility.

As the companies restructure and streamline their facilities, the result will be a reduction in size and fewer jobs. These changes should result in leaner, more profitable enterprises. Consequently, although the jobs will be fewer, they will be far more secure than they were with larger, less competitive plants. This, however, is small consolation to those who lose their jobs.

Employee Stock Ownership Plan

With so many plant closures, the question has been asked as to whether or not some could be reactivated if the employees bought and operated them, as was done with Weirton Steel. This possibility would depend on the individual plant. In the case of Weirton the facilities were competitive, the raw materials were available, and the market for the plant's output was good. Further, Weirton had a fine reputation for making tinplate, its principal product. In addition, the Weirton work force was willing to take a pay cut of 32 percent, which reduced employment costs at the plant by one-third.

It would be necessary for other plants that have been closed down or will be closed down to have at least most of these attributes before an employee stock ownership plan (ESOP) could be considered. Examples will illustrate the point. Most of the Bethlehem Steel plant at Lackawanna was closed in 1983 because it was losing large amounts of money. The company retained a bar mill and a galvanizing line and since the end of 1983 has continued to operate coke ovens. The abandoned facilities, particularly the hot-strip and cold-reduction mills, were of pre–World War II vintage and had not been improved or updated to the extent needed to make them competitive. Lackawanna was not capable of making a quality product at a competitive cost to meet the specifications of the automotive and appliance industries and, as a result, its market had dwindled considerably.

Under these circumstances, it would have been difficult for the employees to raise the money to buy the plant even though Bethlehem might have been willing to sell it at a price below the plant's book value. Also, even though the employees agreed to reduce their wages, this would not have been adequate since the facilities and the market were not conducive to the successful operation of an ESOP at that location.

The Youngstown Sheet and Tube plant at Youngstown, Ohio, which was closed in 1977, was also considered a possibility for an ESOP. This did not come to fruition for it was impossible to raise the funds to convert the plant to a competitive entity. Its blast furnaces and open hearths were obsolete, as was the 1936 vintage strip mill. It could not make the quality steel at competitive costs, and so a complete revision of the plant with an electric furnace and continuous caster was considered. This would have cost much more money than could be raised privately and government help was not forthcoming.

Another basic difference between Weirton and a number of abandoned steel plants throughout the country is that Weirton was not shut down because it was an economic liability. Rather, as indicated elsewhere, National Steel (now National Intergroup) decided to divest itself of Weirton because it wanted to invest its capital funds in projects that would bring a higher return. Thus, although the return on investment from Weirton was low and not satisfactory in terms of National's goals, it still was an economically viable entity. This made it possible for the employees who bought the plant to raise the necessary funds.

In other circumstances in which plants are shut down because they are losing propositions, it would be much more difficult, if not impossible, to raise the funds to buy the plant, even though the company would be willing to sell it to the employees for a sum to be paid in the future when and if the enterprise became profitable. Most steel plants that have been abandoned were shut down either because costs were too high for the facility to be competitive and profitable or because the market had deteriorated. In terms of cutting costs, there are a number of elements—including raw materials, utilities, and labor—that represent possibilities. In the Weirton case, as indicated, labor costs were cut by almost one-third. The reduction in costs, however, can be achieved more readily than the development of new markets, and it is difficult to see how funds could be raised for an ESOP if the market for the plant's products had declined appreciably.

In sum, the possibility of bringing an abandoned plant into operation as an independent company under an ESOP formula depends on the circumstances surrounding each individual plant.

20 Financing New Facilities

The recent depression in the steel industry, which was accompanied by heavy losses and a considerable reduction in cash flow, has brought with it a change in the method of financing costly capital equipment. Traditionally, the money to install equipment for expansion or replacement, such as a blast furnace or a rolling mill, was either available through depreciation charges and retained earnings or, in more recent years, through borrowing.

In the immediate postwar period a number of companies financed the replacement and expansion of facilities through accrued depreciation and retained earnings. However, as the need for more facilities became apparent, both in terms of growth and replacement, and because the cost of these facilities increased sharply due to inflation and technical change, depreciation and retained earnings were by no means adequate to the task. As a result the companies were forced to increase their debt. This doubled for the industry between 1960 and 1970, increasing from $2.488 billion to $5.134 billion. By 1980 it had almost doubled again, as it rose to $9.803 billion. Further, interest charges increased significantly from $101 million in 1960 to $288 million in 1970 and to $873 million in 1980.

Clearly, the increased debt and the accompanying interest represented an onerous burden on the companies, and in 1982 and 1983, with profits turning to large losses, the amount of internal funds (as well as the ability and willingness to borrow) had diminished sharply. Nevertheless, for the companies to attain a competitive status, there was need to invest in new facilities, such as continuous casting. Faced with these circumstances, various types of off-balance-sheet financing and other arrangements have been developed.

One of the major moves in capital investment was the arrangement made in 1981 by USS Corporation in financing the new seamless-pipe mill and accompanying continuous caster at the Fairfield Works in Alabama. A group of oil companies agreed to pay a specified semiannual amount beginning in 1984 for pipe to be taken either immediately or when needed. This sum will be paid to a group of financing institutions who have put up some $540 million that constitutes the bulk of the $700 million investment.

In 1983 Bethlehem Steel Corporation announced the construction of two continuous casters to be financed off the balance sheet. Details of this financing are given elsewhere in this book in the chapter dealing with Bethlehem Steel. As indicated there, Bethlehem leased the equipment from financial institutions that provided funds and will begin payments on a tonnage basis when steel is produced. Inland Steel financed two continuous-casting machines in a similar manner, and USS Corporation has two additional continuous-casting machines that will probably be financed with an off-balance-sheet arrangement, where no money is paid until the unit goes into production.

All the major suppliers of equipment to the steel industry recognize that it is now possible to sell their equipment only if they offer suitable financing terms, and in most instances, it will be off-balance-sheet financing with the company using the facility under terms of a lease. This method is not confined to the United States but is applied on a worldwide basis.

Steel-mill equipment financing is also arranged in a more conventional manner but on very favorable interest terms. For example, the new integrated steel plant to be built in South Korea has a number of facilities that have been purchased from various equipment producers. These include the steelmaking shop and sinter plant which will be provided by Voest of Austria on the basis of an 11½-year payout with interest at 6.75 percent. The coke ovens are to be furnished by Otto, a West German company, with an eight-year payout and interest at 6.95 percent. The blast furnace will be built by Davy, a British company, on a 9½-year payout at 6 percent interest. The continuous-casting unit was awarded to Mannesmann with 6.95 percent interest and a 9½-year payout. The hot strip mill will be furnished by Mitsubishi of Japan with a 9½-year payout at 6.95 percent interest.

The willingness of mill suppliers to provide such low-cost financing (usually from their governments through an export–import bank or its equivalent) is based on the lack of business and the desire to maintain their organizations and employment. Government agencies in these countries have cooperated in this endeavor. The interest rates are relatively low (because South Korea is a developing country) and the governments of the various countries supplying the financing take this into account. Such low interest rates theoretically would not be made available to industrialized countries.

During the next few years low-cost financing will be sought by steel companies throughout the world to enable them to install large capital facilities. This is even true of joint ventures with more than one company participating. For example, the recently announced electrolytic galvanizing line to be jointventured by Inland and Bethlehem will most probably be financed by a third party and leased to Bethlehem and Inland. Thus the mill builders must practically be in the banking business in order to sell their equipment. However, since there is a risk on the lender's part, the willingness to provide attractive financing will depend on the financial stability of the steel company involved.

Productivity and New Technology

Restructuring must include the adoption of new technology and the installation of facilities to put it into practice. This will increase productivity and improve the competitive position of the company, since improved equipment will reduce the required man-hours per ton and effect substantial cost savings. The current productivity level for the steel industry in the United States is in the area of 6 to 7 man-hours per ton shipped. If this could be reduced to 4 or 5 man-hours, it would cut labor costs per ton by $42 to $44.

During the last quarter century new technology in steel has been directed toward achieving continuous operation in the production of steel to replace the batch processes in the various steps of steelmaking that have existed for more than a century. Installations such as continuous casting, continuous-annealing lines, and continuous-coating lines are examples of this effort. By far the most outstanding development in this respect in the post-war period has been continuous casting.

Continuous Casting

This process bypasses several steps in the conventional production of steel, eliminating the pouring of steel into ingot molds, stripping the molds from the ingots, placing the ingots in soaking pits to develop an even temperature, and finally, the primary rolling stage by which the ingot is rolled into semifinished form, either a slab, billet, or bloom. The elimination of these steps, particularly the rolling operation, has saved energy and increased productivity and improved the yield from raw steel to finished product by at least 10 percent.

In the continuous-casting process, the steel is tapped from the furnace into a ladle and then poured directly into the continuous caster. It solidifies as it passes through and emerges as a slab, billet, or bloom. Steel from continuous casting in some ways is superior to steel produced by the conventional ingot-mold method, particularly in surface quality.

During the past fifteen years, continuous casting has been installed by the steel industry throughout the world in ever increasing tonnages. In 1969, only 3 percent of the steel produced worldwide was continuously cast. In 1976, this figure rose to 17 percent. By 1980, it had grown to 30 percent and by 1983 to approximately 45 percent. The leading country in terms of steel continuously cast is Japan, with approximately 80 percent of its 1982 production processed through casters. The EEC has increased its use of the process so that by 1982 53 percent of its steel was continuously cast.

The U.S. industry was slow to adopt this process, having only 11 percent of its steel continuously cast in 1976 as opposed to 35 percent in Japan for that year. By 1981 the percentage in the United States had risen to 22, and this increased sharply to 31 percent in 1983. Several units were installed in 1983 throughout the U.S. industry, and a number of others are currently under construction. This will raise the percentage of steel continuously cast significantly in 1984 and 1985. Currently in the entire U.S. steel industry, there are 121 continuous casting machines with a potential productive capacity of some 58.0 million tons. Many of these are operated by the so-called minimills while a number, particularly the slab casters, are in operation in integrated steel plants. Virtually every integrated company has a continuous caster either recently completed, under construction, or firmly committed.

USS has just installed a unit at Lorain, Ohio, and has two more committed: one for Gary, Indiana, and one for Fairfield, Alabama. Bethlehem has just completed a unit at its Steelton plant and has two more committed, one for Sparrows Point and another for Burns Harbor. Republic has just completed a large slab caster at its Cleveland Works. Inland Steel (in addition to its existing units) has another under construction. Armco has just completed a caster at its Ashland plant. J&L has just completed a major slab caster at its Indiana Harbor Works, and in the last two years, National has installed a continuous caster at its Granite City, Illinois, plant. Wheeling–Pittsburgh, in the last two years, has completed two units, one at Steubenville, Ohio, and the other at Monessen, Pennsylvania. Rouge Steel has announced its intention to install a slab caster at its plant in Dearborn, Michigan.

Many of these facilities are large by any standards and several are capable of processing 3.0 million tons of steel annually. As a consequence, the 1984 ratio of continuous cast steel could be close to 40 percent, and by 1986 it will be at least 50 percent of the industry's steel output. Continuous casting is considered a necessity by the U.S. producers and there is little doubt that even more units than those currently planned will be installed in the remaining years of the 1980s. By 1990, it is quite reasonable to assume that the industry's capacity to process steel through continuous casters will be as high as 75 percent of its output.

By comparison with the integrated plants, the electric-furnace operations, both large and small, have a high percentage of their output continuously cast. The vast majority of minimills employ continuous casting for 100 percent of their raw steel output. The large electric-furnace operations, such as Northwestern Steel and Wire, Lukens, and Laclede, also have a substantial percentage of their steel continuously cast.

Currently there is need for growth in this area, since the process is a decided help toward improving productivity by providing a higher yield from raw steel to finished product and also eliminating several stages in the conventional production of steel, as well as reducing energy requirements. The increased yield tends to compensate to some extent for the reduction in steelmaking capacity, since the industry will be able to ship substantially more tonnage from its raw-steel output as the percentage of continuously cast steel increases.

Continuous Annealing

Another technological advance which has eliminated the batch process is continuous annealing for the production of steel sheets. This process has been in operation for many years for the production of tinplate; however, it has only recently been applied to sheets. The two installations now operating in the United States are at the Inland Steel plant and the Burns Harbor plant of Bethlehem Steel. The process provides a higher-quality steel, particularly for difficult automobile applications, and within the next two to three years, other integrated companies serving the automotive industry will have to install this equipment if they are to remain competitive. As previously indicated, this process results in a literally enormous reduction of time as well as a significant improvement in quality by comparison with batch annealing. The entire batch-annealing process, including cooling, requires three to four days, whereas with a continuous-annealing line, the process is completed in less than an hour.

High-Stability Coke

Another development, which is not particularly new but has recently become more widespread and, hopefully will become standard throughout the industry, is the production of coke with a higher stability factor. This requires more extensive pulverization of coal as well as a longer coking time and, in a number of instances, results in additional cost for coke production. However, the coke performance in the blast furnace more than makes up for this through the reduction in pig-iron costs. The higher stability coke is needed in the larger blast furnaces, and the industry is expanding the size of its furnaces whenever possible.

Blast Furnace

In the blast-furnace segment of the industry, coal injection, and external desulfurization of iron in the ladle, along with oxygen enrichment and a more intensive burden preparation, have improved productivity considerably. With only a small increase in size, furnaces that produced 1500 tons per day a decade or two ago now produce double that amount. However, even this can be increased with improved stoves, permitting higher blast temperatures as well as other items mentioned above. Several blast furnaces, which are being reconditioned at the present time, will operate at higher blast temperatures and be able to increase output from 3200 tons per day to 4000 tons per day. This will increase productivity.

Ladle Metallurgy

Ladle metallurgy is one of the latest technological advances whereby much of the steel refining is performed in the ladle after it has been poured from the furnace. This can consist of a variety of operations, including stirring the bath with gases such as argon, and reducing the sulfur content so that the resultant steel is much cleaner. The most advanced form calls for a ladle into which an electrode is inserted to maintain heat while the refining process is carried out. The result is a cleaner steel with the undesirable elements much reduced. Currently, there is a great deal of discussion about the installation of ladle metallurgy, although very little has been done. Most steel operators recognize its benefits and are pressing their managements for funds to install it. The resultant product will make the company using it more competitive in terms of steel quality.

Electrolytic Galvanizing Line

The demand for galvanized steel for automobiles has been increasing and will continue to increase. It will be used extensively in the automobile body to prevent corrosion. As a consequence, there is a demand for this product that very soon will exceed the supply unless more facilities are installed. Bethlehem and Inland have announced the formation of a joint venture to install an electrolytic galvanizing line, and National will probably make a similar arrangement. The latter line will be installed at the Great Lakes plant of National Steel to serve the automobile industry. Rouge Steel is also considering the construction of a line by 1985. More electrolytic galvanizing lines will be needed in the decade to come; however, there is the danger that too many lines will be installed, resulting in overcapacity.

The foregoing technological advances are available currently. In addition to these, there are a number that will be used in the future despite the fact that at present they are in the experimental stage. In respect to these developments, it would be well for the U.S. industry to use them wherever possible (at least experimentally) so that the companies will be in the forefront of technological development and not, as has been the case in many instances, in a position of catching up. Some of these advances include direct rolling of cast slabs without reheating, the use of laser beams for the analysis of metal, horizontal continuous casting for small heats of steel, and low-temperature rolling to conserve energy.

22 United States Steel Industry's Ability to Satisfy Domestic Demands

The ability of the U.S. steel industry to satisfy future domestic demands must be evaluated in terms of its projected capacity as well as an estimate of the economy's future steel requirements.

Steel Supply

According to the present plans of a number of steel companies raw steel capacity throughout the nation will shrink to 125 million to 130 million net tons by 1985. This will reduce the capability of the industry to ship steel, although the additional installation of continuous casting units will neutralize the cut to some extent. By 1986 some 50 percent of the steel produced will be continuously cast. This is in contrast to 20 percent in 1980 and 31 percent in 1983. As a result the ability of the industry to ship steel will be in the area of 90 to 95 million tons by 1985.

As far as individual products are concerned, capacities will vary substantially depending on finishing facilities. In the past and, without question, in the future, light flat-rolled products will dominate the steel industry's shipments. In 1979 (the last year in which 100 million tons of steel were shipped) light, flat-rolled products, including all types of sheet and strip as well as tin-mill products, accounted for some 48 percent or 48 million tons of total steel shipments. In 1983, when shipments fell to a low of 61.6 million tons, light, flat-rolled products accounted for 49 percent, or 30 million tons, of steel shipments.

On the assumption that the prosperous conditions of 1979 were repeated—bringing with them a demand for light flat-rolled products equivalent to that year—current hot-strip mill capacity, which exceeds 65.0 million tons, would be more than enough to roll these products. Approximately 45.0 million tons of this capacity are represented by twelve second-generation, fully competitive hot-strip mills installed in the 1960s and early 1970s. Thus it is evident that the capacity to produce sheets, strip, and tin-mill products would be adequate to satisfy any demand in the future. This, however, takes into account only finishing facilities. There is a question as to whether or not there will be enough raw steel to put through them, considering the demands for other products. In respect to most of the other products, there will be enough finishing facilities to produce them, because in addition to strip mills there are enough plate, pipe, bar, and structural mills to cover most of the needs of the country for each of these products. However, here again, the ability of these mills to produce depends on the total availability of raw steel.

Steel Demand

The heavy demand for steel is concentrated in relatively few industries. The principal consumers are the automotive, construction, railroad, oil and gas drilling, machinery, and container industries.

The automotive industry is the largest single consumer of direct-mill shipments as well as the principal consumer of light-flat-rolled products. In years such as 1973, 1977, and 1978, when record numbers of motor vehicles were produced, the automotive industry consumed very large quantities of steel. For example, in 1973 motor vehicle production reached 12.7 million units consisting of 9.7 million passenger cars and 3.0 million trucks. In that year, the automotive industry took 23.2 million tons of steel in direct shipments, and it is estimated that this should be augmented by at least another million tons which would include shipments to the industry of other steel products and from other sources. This would include fasteners and steel from steel service centers.

In 1977, when motor vehicle production again reached 12.7 million units (with 9.2 million passenger cars and 3.5 million trucks) the automotive industry had direct shipments from the steel industry of 21.5 million tons, to which another million tons could be added. Thus, there was a decline of approximately 2.0 million tons for the production of the same number of vehicles, which indicated the effect of the downsized vehicles on steel consumption.

In 1978, automotive production reached an all-time high of 12.9 million units—9.2 million passenger cars and 3.7 million trucks. In that year, direct shipments to the automotive industry were 18.6 million tons, indicating a further shrinkage in the size of the automobile and the requirement of less steel.

In 1980 through 1982, the automotive industry suffered a severe setback, as total vehicle production fell to 8.0 million units for 1980 and 1981 and 7.0 million units in 1982.

In 1982, with a sharp drop in production, direct shipments of steel to the automotive industry fell to a low of 9.3 million tons, which was 15 percent of the steel shipments as compared with 21.0 percent in 1973 and 21.7 percent in 1979.

In 1983 automotive production revived to 9.2 million vehicles—a significant increase. However, steel shipments recovered to only 12.1 million tons. If we compare 1983 with 1964, when approximately the same number of vehicles were produced, steel shipments dropped sharply. In 1964, steel shipments were 18.4 million tons, or more than 6.0 million tons over the 1983 figure for the same production. Steel imports also were a factor.

In addition to the downsizing of cars, imports of automobiles have taken as much as 27 percent of the U.S. automotive market. Thus, it is questionable whether or not the automotive industry will again reach a production near 13.0 million motor vehicles. Imports which, in the three-year period 1980–1982 accounted for an average of 27 percent of the American market, are up sharply from the level of the early 1970s. In those years, passenger car imports were in the 1.5-million vehicle range; by 1980 to 1982 they had reached the 2.0-million to 2.4-million range. Truck imports in the early 1970s were in the 200,000-vehicle range but by the early 1980s they were more than 400,000 annually. Over 70 percent of these imports in the last few years came from Japan.

In response to pressure from the United States, the Japanese agreed to restrict their exports to approximately 1.68 million passenger cars per year for a three-year period, 1981 to 1983. Recently this agreement has been revised upward to 1.8 million passenger cars for the year 1984.

With imports and smaller cars, the automobile industry as a market for steel will be as much as 8.0 to 10.0 million tons less than it was in the record years of the 1970s. The downsized car requires an average of 1.1 tons of steel as opposed to 2.0 tons in 1973.

Another area where steel markets have shrunk is in tinplate and tin-mill products for cans and containers. The market fell from a high of 7.5 million tons in 1974 to 4.3 million tons in 1983, a loss of 3.2 million tons. These products, because of their end-use, are considered the least susceptible to cyclical fluctuation in the economy. Therefore, the drop cannot be attributed to generally depressed conditions; the loss of market is due to inroads of other materials. Aluminum, for example, has captured the beer can market, which consists of 20 percent of the entire can market, while other materials, such as plastics and impregnated paper, have also made a dent in the demand for tinplate.

The drop in steel shipments to the automotive and canning industries could represent a cut in demand of 11.0 to 13.0 million tons from the peak. These are permanent losses that will increase, as tinplate production is expected to decline 1 percent to 2 percent per year for the remainder of the decade.

Other products, such as structural steel and plates, which are sold in great part to the construction and machinery industries, can be expected to have a revival since these industries move with the economy. In the matter of oil-country goods, where shipments amounted to 4.2 million tons in 1981, there has been a precipitous decline to less than 0.7 million tons in

1983. The 1981 figure was the result of a boom year when oil drillers were operating 4,530 rigs and leasing many of them at a rate of $12,000 to $14,000 per day. Therefore, none of the drillers wanted to see the rigs idle for lack of pipe. Consequently, huge inventories were built up, so that in early 1983 there were some 5.0 million tons on hand in the oil patch. Even though drilling should pick up again, it is doubtful that the 1981 demand will be repeated; it will probably drop back to an average of 3.0 million tons of steel per year.

Another steel market that has experienced a permanent decline is railroad rails. Once a most important steel consumer of several million tons per year, the railroad industry demand has fallen to an average of about 1.0 million tons of rails per year for the last ten years. In 1982, however, the figure dropped below 500,000 tons since this was a particularly bad year. The recovery in 1983 was modest, as shipments were about 610,000 tons. The future does not seem particularly bright since the railroads are consolidating and abandoning thousands of miles of track. In this process rail is taken up and, if it is in good condition, stored for future use. Consequently, it appears that rail shipments will continue into the foreseeable future at under 1.0 million tons.

From the foregoing it is evident that there have been some permanent losses in the steel market; however, the market still remains substantial. Projections of domestic shipments for the future will probably not top 90 million to 93 million tons for the remainder of the decade. Imports will range from 16 million to 17 million tons, due in great part to the quota arrangements with the Western Europeans and the self-restraint of the Japanese. The total U.S. market in the best year during the remainder of the 1980s will probably be no more than 107 to 108 million tons, of which the U.S. steel industry will be capable of supplying 93 million tons, or about 85 percent to 86 percent.

The reduction in the steel market corresponds to some extent with the reduction in steelmaking capacity. However, when demand falls below peak levels the competition for the steel market will be intense. In 1982 and 1983, with a record-low steel demand, competition among the domestic producers as well as with the importers was extremely keen. Discounts from the list price were the rule rather than the exception; in fact, it was rare during this period that any company received the list price for its product. These discounts were maintained on steel sheets even after the market demand for steel sheets strengthened with the revival of the automotive and appliance industries. During 1983 discounts on cold-rolled sheets were as high as $80 to $90 below the list price of approximately $500. About the same range pertained to hot-rolled sheets, which listed for approximately $400 and sold for a price very close to $300. Discounts on plates were well over $100 per ton. As the market improved discounts were reduced; however, they were still subtantial for most products in the first part of 1984.

Price competition will remain except for those periods when the industry operates at better than 90 percent of its capacity. This was the case in 1973 and 1974. However, in virtually every year when the operating rate dropped below 85 percent, there was price competition among the domestic steel producers. This is a fact that has not been generally recognized, principally because prices are published in newspapers and magazines, such as *American Metal Market* and *Iron Age*, and most people outside the industry believe that all companies receive these prices. A number of observers have stated that competition in steel is a matter of quality and service but not price. This is not a fact except for the boom years. In the years immediately ahead, there is little doubt that price competition will continue to be vigorous.

23 Conclusions

The steel companies in the United States are currently in a critical situation where losses have been so great that many are justly concerned about survival. Clearly, action had to be taken to cut losses and hopefully, with an upturn in the market, insure a return to a profitable position. There have been a number of changes and significant restructuring in the last two years and more will be forthcoming in 1984 and 1985 so that the industry will be considerably different in the years ahead.

Some of the following conditions under which the steel companies will operate and the problems they will face are familiar, while others present a new challenge.

1. The industry will be smaller in terms of total raw-steel capacity, which will be approximately 125 million net tons in 1986 and by 1990 could shrink below 120 million net tons with the elimination of some open-hearth capacity and quite possibly one or two BOF shops.
2. With the installation, between 1983 and 1986, of some fifteen new large-tonnage continuous casters, the portion of steel continuously cast will be in excess of 50 percent by late 1986. In the early 1990s, because of competition and the economies to be derived from continuous casting, the figure could well reach 75 percent of the steel produced.
3. No new blast furnaces will be built either as additions or replacements before 1990. It is quite probable that when existing blast furnaces have to be relined and renovated, decisions will be made to abandon some units. This is due to the high cost involved, which ranges from $15 million to $40 million depending on how much is to be done. There are at least eighteen furnaces on which this decision will be made in the next two years. Those that remain and are relined will be improved and to some extent enlarged to increase their capacity and efficiency.
4. There will be some growth in electric-furnace capacity over the present figure of 48 million tons to a maximum of 55 to 57 million tons by 1990. Some of this tonnage will replace abandoned blast furnaces.

   The electric furnaces will constitute almost one-half of the steelmaking capacity in the country during the next decade. However, the scrap problem—in terms of quality—must be solved before this expansion can take place.

5. Because of the high cost of new and revamped facilities, capital investment will be confined to essential items such as continuous casting and those other facilities necessary for a company to reduce costs, save enery, and maintain operations as well as its competitive position. There are limitations, however, to the capital investment of even the largest steel companies when they are incurring continual losses. Profitability must be restored for any extensive capital investment program. New technology will be installed if it promises to increase operating efficiency and reduce costs. An example is hot charging and direct rolling, which should reduce man-hours by a significant amount.
6. There will be a continuation of the trend among minimills to increase in size and produce more sophisticated products. However, there are limitations on how far this can go. The finishing facilities needed for such items as wide sheets, cold-reduced sheets, heavy structurals, tinplate, and other items are too expensive for minimills. There is also a danger here. As the plants grow larger, they will lose some of the advantages they enjoyed as small compact units.
7. New types of financing, much of which is off-balance-sheet, will continue. However, financing institutions that own and lease equipment to the steel companies may experience a degree of risk that may restrict this type of investment in the future. Economically unstable companies will not be able to obtain this kind of financing.
8. More obsolete equipment will be scrapped or replaced so that by 1987 or 1988 very little will be in operation. The result will be that most remaining plants will be efficient units.
9. Many integrated plants will shrink in size and restrict output to those products that can be made profitably. This applies to those companies with several integrated plants, such as the new LTV Steel Corporation and USS. This approach is also being pursued overseas, where Nippon Steel has designated four integrated plants as specialty producers.
10. Employment opportunities, which have declined in the past decade, will continue to be limited. Since 1973, the number of employees has dropped from 500,000 to less than 300,000, or more than 40 percent. Currently there are a number of employees on temporary layoff, but 200,000 jobs have been permanently eliminated. As more facilities are closed, there will be a further loss of employment. However, this should stabilize within the next year since most of the obsolete facilities have already been abandoned.

    More wage and benefit concessions will be sought by management but these will be limited to individual companies, plants, or product lines that are unprofitable. Work-rule changes will also be sought.
11. Imports will continue to take a significant share of the U.S. market. Attempts are being made to limit them to 15 percent. If these attempts

fail, import market penetration could be as high as 20 percent to 23 percent, which will create serious problems for a number of companies (particularly if the incoming steel is subsidized and sold at prices below its cost of production). If this occurs, it could result in the closure of more U.S. capacity. This, in turn, would increase the dependence of the U.S. consumers on foreign steel. If this steel comes in as a supplement to domestic supply, rather than in competition with it, prices will increase sharply and remain high.

12. The Third World with its growing steel industry will be looking for an increased share of the U.S. market. Political and economical pressures will be brought to bear to permit countries in the Third World to ship more steel into the United States. Except for Mexico's offer of self restraint, there are no present limits on Third World imports except for individual suits in connection with countervailing duties or dumping. However, other offers are expected.

    The restrictions proposed under the 15 percent global quota indicate that the Third World and Canada will have only 5 percent of the U.S market with 10 percent divided between the European Community and Japan. The EEC currently has slightly more than 5 percent and the Japanese want parity with the EEC. This situation will be sharply contested and could result in negotiations which would increase the global quota above 15 percent.

13. Some attempts at mergers will be made involving the large integrated companies. If the Justice Department changes its criterion for measuring concentration, mergers can take place between large steel companies. However, if the Justice Department fails to take into account imports and substitute materials in calculating concentration, mergers may be limited to smaller companies or possibly to larger and smaller companies.

    There is a possibility that foreign steel companies may either take over or form partnerships with steel companies or parts of steel companies in the United States, such as the recent arrangement between Wheeling–Pittsburgh and Nisshin Steel of Japan. There are a number of facilities that could readily be acquired in this manner, including the Gadsden Works of Republic Steel which must be sold, as well as the steel plants of National Intergroup. The former plant could be operated as a finishing facility dependent on outside slabs.

    The attempt of USS to form an alliance with British Steel Corporation will be duplicated by other companies, particularly those who wish to take advantage of lower-cost steel made in the Third World. To obtain a permanent source of this steel, some type of liaison between U.S. companies and Third World companies based on investment is a possibility. The Japanese are also prospects for such arrangements considering the venture between Wheeling-Pittsburgh and Nisshin and proposed National-Nippon Kokan agreement. However, to anticipate a rash of Japanese investments in the United States is not realistic, since many Japanese companies are reducing their capacity and diversifying.

14. The market for steel products, particularly sheets, oil-country tubular goods, plates, and structurals will almost certainly not reach the peaks achieved in the 1970s. As a consequence, competition (including price competition) will be strong not only with imported steel but among the domestic producers as well. In other than boom markets, most of the steel companies will be competing for business in terms of quality, service, and price.
15. The restructuring of the steel industry in the United States is taking place in a global context. Many of the changes in respect to trade among steel countries have taken place gradually during the past 30 years. World trade has increased dramatically from 10 percent of steel produced in 1950 to more than 25 percent of the production in the late 1970s and early 1980s. Actual tonnage grew from 10 million to 135 million metric tons in raw-steel equivalent. This growth in trade was made possible since steel producers found markets not only in areas where little or no steel was made but in those areas where large tonnages of steel had been produced for many years. The largest, most sought after market for most steelmakers is the United States, which up to 1973 was the largest steel producer in the world.

The U.S. steel industry has experienced strong competition within its home market, first from Western Europe and Japan, but more recently from Third World countries. Consequently, it has been and will continue to be incumbent on producers in the United States to meet this competition by improving operations whenever possible. Further, when this competition is considered unfair due to subsidies, it will be necessary to have the government remedy the situation with appropriate action.

In the early 1980s, with a depressed steel market, it was evident that there had to be radical surgery as far as steel facilities were concerned to permit companies to survive in a competitive struggle intensified by the decline in the steel market. However, the restructuring which the U.S. steel companies have undergone and will undergo in the next year or two has also been carried out in other countries so that they will improve their respective competitive positions.

The Common Market industry in Western Europe has cut back its capacity sharply since 1980. It recognizes the need to reduce and eliminate capacity that most likely will not be needed in the future and to improve what remains with the adoption of the latest technology. The Japanese, who have been given credit for the most modern steel industry in the world, also face the challenge of reducing and restructuring their companies. Compared with 1980, the European Community industry will be some 30 million tons smaller by 1986, when one considers the closures that have taken place and are planned by 1986. Further, there will most probably be mergers among companies, although to date some of these efforts, in spite of government encouragement, have failed to materialize. This is particularly true in West Germany. Joint ventures involving companies in Luxembourg, the Netherlands, and Belgium have been proposed.

The Japanese companies have reduced their functioning capacity by abandoning a number of facilities and have signified their intention to extend this process. Many of the facilities will not be officially abandoned but placed in mothballs with little likelihood of reactivation. In 1980, the theoretical, nominal capacity of all the plants in Japan was 150 million metric tons. By 1984, the reliable operative capacity is much closer to 130 million tons and most probably will decline to 120 million tons. The Fukuyama plant of Nippon Kokan, once the world's largest with 16 million metric tons of raw-steel capacity, will be reduced to less than 7 million tons in terms of active reliable steelmaking potential. This does not mean that there will be wholesale demolition of facilities but rather a mothballing operation. Nippon Steel has closed several parts of plants, reducing its capacity from 47 million tons to less than 35 million.

The steel companies in almost all of the industrialized countries are striving to improve their competitive position; thus, in anything less than a boom year, competition will be severe and this gives the U.S. companies a further incentive to restructure. In contrast to the industrialized countries, the Third World has increased its steelmaking facilities but with the exception of one or two countries, at a very slow rate. The plants that have been constructed are new and the costs of production, particularly labor costs, are low, thus giving a decided advantage to those operations in South Korea, Taiwan, and Brazil.

On balance, in spite of Third World growth, the steel industry outside of the Soviet orbit has declined and will continue to decline through 1986 and 1987. It will, however, be an improved, more competitive industry as companies abandon obsolete facilities and put their capital investment in the plants that are left. With these developments, competition will be impossible in the near future for any company in the industrialized world that has not undergone some form of restructuring. Another consequence of the elimination of obsolete and marginal melting facilities may well be a greater demand for and trade in semifinished steel, a continual supply of which for some companies constitutes their life's blood. Kaiser Steel is an example. Having shut down its melting facilities, it must obtain semifinished steel to function. However, the supply of this material on a long-term basis will be limited. Further, both the 201 Petition and the Trade Bill, which call for global quotas, include semifinished steel.

In regard to global quotas, whatever percentage of the U.S. market is conceded, they have one definite advantage. All parties involved, American and others, know just how much steel can be imported. This will eliminate or greatly reduce the suits on dumping and countervailing duties, which generate a climate of uncertainty. They are also expensive and time-consuming for those who file as well as for those who must respond.

A global quota is not meant to exclude imports but to restrain excessive imports which have led in the past to destructive competition and could do so in the future. The United States needs imports since the domestic industry cannot satisfy the full requirements of the economy in a reasonably good year. A 15 to 17 percent quota would allow these needed imports since the U.S. steel industry, even with its reduced capacity, is capable of serving 85 percent of the market.

The demand for steel on a global basis for which there will be keen competition comes currently and will continue to come from two sources: growth and replacement. The industrialized countries are faced principally with a replacement market, while the Third World can look to growth.

The replacement market is huge in terms of the steel involved in consumer durables such as motor vehicles, as well as machinery and industrial equipment that wears out or becomes technologically obsolete. Of great significance is the infrastructure which must be expanded and replaced. In the United States, for example, some 250,000 bridges are either physically or functionally obsolete and must be replaced.

The Third World is wide open in terms of growth potential. Basic infrastructure must be put in place; industrial equipment is needed for new plants; and ultimately consumer durables such as automobiles will be in strong demand. The basic problem, however, is the Third World's ability to pay for the steel required for these applications.

The demand for steel on a worldwide basis will be satisfied by more efficient plants with higher productivity. This improvement will be accomplished by restructuring and updating plants in terms of corporate efficiency and new technology, not once but continuously.

The principal cause for the heavy losses incurred by steel companies in 1982 and 1983 was the sharp decline in demand for steel in most consuming industries. A revival in the steel market which seems to be occurring in 1984 is necessary to restore profitability. The restructuring is essential; however, only so much can be accomplished if the demand for steel is weak.

Bibliography

American Institute of Mining, Metallurgical and Petroleum Engineers (AIME). *Energy Use and Conservation in the Metal Industry.* New York: AIME, 1975.

American Iron and Steel Institute. *Annual Statistical Reports.* Washington, D.C.: AISI, various years.

______ . *Directory of Iron and Steel Works of the United States and Canada.* Washington, D.C.: AISI, various years.

American Metal Market Co. *Metal Statistics: The Purchasing Guide of the Metal Industries.* New York: American Metal Market, various years.

Barnett, Donald F., and Scorsch, Louis. *Steel: Upheaval in a Basic Industry.* Cambridge, Mass.: Ballinger Publishing Company, 1983.

Boylan, Myles. *Economic Effects of Scale Increase in the Steel Industry.* New York: Praeger Publishers, 1976.

Cockerill, Anthony. *The Steel Industry: International Comparisons of Industrial Structure and Performance.* New York: Cambridge University Press, 1974.

Comptroller General of the United States. *Report to Congress of the United States: New Strategy Required for Aiding Distressed Steel Industry.* Washington, D.C.: General Accounting Office, 1981.

Cordero, Raymond, and Serjeantson, Richard, eds. *Iron and Steel Works of the World,* 7th ed. London: Metal Bulletin Books, Ltd., 1978.

Crandall, Robert W. *The U.S. Steel Industry in Recurrent Crisis.* Washington, D.C.: The Brookings Institution, 1981.

Deily, Richard L., and Pietrucha, William E. *Steel Industry in Brief: Databook, U.S.A.* Green Brook, N.J.: Institute for Iron and Steel Studies, 1983.

Gold, Bela; Rosegger, Gerhard; and Boylan, Myles G. *Evaluating Technological Innovations.* Lexington, Mass.: Lexington Books, D.C. Heath & Co., 1980.

Greenhouse, Steven. "Analysts Hail Republic–LTV Link," *New York Times* (September 28, 1983):D1.

______ . "No U.S. Bar Seen to LTV Steel Plan," *New York Times* (October 3, 1983):D1.

Hirschhorn, Joel S. "Continuing Success for United States Mini Mills" *Metal Bulletin's Second International Mini Mills Conference.* Vienna: March 7–9, 1982.

Hogan, William T. *World Steel in the 1980s: A Case of Survival.* Lexington, Mass.: Lexington Books, D.C. Heath & Co., 1983.

Hogan, William T., and Koelble, Frank T. *Direct Reduction as an Ironmaking Alternative in the United States.* Washington, D.C.: U.S. Department of Commerce, 1981.

Innace, Joseph J. "Independence Day at Weirton," *33 Metal Producing* 22 (February 1984):53–54.

______ . "Reshaping Domestic Steel," *33 Metal Producing* 21 (November 1983):47–52.

International Iron and Steel Institute, Committee on Statistics. *Steel Statistical Yearbook.* Brussels: IISI, various years.

International Iron and Steel Institute, Committee on Technology. *Energy and the Steel Industry.* Brussels: IISI, 1982.

Labee, Charles J., and Samways, Norman L. "Development in the Iron and Steel Industry, U.S. and Canada—1982," *Iron and Steel Engineer* 60 (February 1983):D1–D24.

______ . "Developments in the Iron and Steel Industry, U.S. and Canada—1983," *Iron and Steel Engineer* 61 (February 1984):D1–D26.

LaRue, Gloria T. "U.S. Steel Cuts Bar, Wire, Rod," *American Metal Market* (December 29, 1983):1, 11.

McManus, George. "Electric Furnace Report: Growth That Won't Stop," *Iron Age* 227 (January 2, 1984):107–109.

Marcus, Peter F. *Internationalization of Steel.* New York: Mitchell Hutchins, Inc., March 1976.

Mazel, Joseph L. "Technology Investment Needed to Spur Steel's Survival and Revival," *33 Metal Producing* 21 (October 1983):64–65.

O'Boyle, Thomas F. "U.S. Steel, in Major Revamp, Is Seen Closing Plants; Write-Off Could Exceed $1 Billion," *Wall Street Journal* (December 22, 1983):3.

Serjeantson, Richard; Cordero, Raymond; and Cooke, Henry, eds. *Iron and Steel Works of the World,* 8th ed. London: Metal Bulletin Books, Ltd., 1983.

33 Magazine: World Steel Industry Data Handbook, Vol. 5, U.S.A. New York: McGraw-Hill, Inc., 1982.

U.S. Congress, House of Representatives. *Crisis in the Steel Industry: An Introduction and the Steel Industry in Transition.* 97th Congress, Second Session, Committee Print 97-EE. Washington, D.C.: U.S. Government Printing Office, 1982.

U.S. Congress, Office of Technology Assessment. *Technology and Steel Industry Competitiveness.* Washington, D.C.: U.S. Government Printing Office, 1980.

U.S. Steel Industry Task Force. "A Comprehensive Program for the Steel Industry," *Report to the President of the United States.* Washington, D.C.: December 13, 1977.

Index

Index

About the Author

Rev. William T. Hogan, S.J., was graduated from Fordham College in 1939 and received the M.A. and the Ph.D. in economics from Fordham University in 1940 and 1948.

He has conducted economic studies of the steel industry and other basic, heavy industries for the past thirty years. During this time, he has authored a number of books, including *Productivity in the Blast-Furnace and Open-Hearth Segments of the Steel Industry,* the first detailed study on the subject of steel productivity; *The Development of American Heavy Industry in the Twentieth Century;* and *Depreciation Policies and Resultant Problems* (1967).

His five-volume work, *Economic History of the Iron and Steel Industry in the United States* (Lexington Books, 1971) covers industry developments from 1860 to 1971. The first two companion studies to this are: *The 1970s: Critical Years for Steel* (1972) and *World Steel in the 1980s: A Case of Survival* (1983).

In 1950, Father Hogan inaugurated Fordham University's Industrial Economics Research Institute, which has produced numerous studies dealing with economic problems of an industrial nature. He has appeared before legislative committees of both the U.S. Senate and the House of Representatives and has testified several times before the Ways and Means Committee of the House on legislation affecting depreciation charges and capital investment. He has also appeared before the Finance Committee of the Senate to testify on tax incentives for capital spending. He was a member of the Presidential Task Force on Business Taxation and a consultant to the Council of Economic Advisers to the President and the U.S. Department of Commerce.

During the past fifteen years, Father Hogan has visited most of the steel-producing facilities in the world and has delivered papers at steel conferences in France, the United Kingdom, Switzerland, Sweden, Czechoslovakia, Russia, Venezuela, Brazil, South Africa, India, the Philippines, and Japan. He is the author of numerous articles on various aspects of steel-industry economics.